Technology, Programming, and Applications in Industrial Robotics

Technology, Programming, and Applications in Industrial Robotics

Contributors :
Carlos Rodriguez-Donate,
Luis Morales-Velazquez, *et al.*

AURIS REFERENCE LTD.
London, UK

Technology, Programming, and Applications in Industrial Robotics
Contributors : Carlos Rodriguez-Donate *and* Luis Morales-Velazquez, *et al.*

Auris Reference Ltd., UK

www.aurisreference.com

United Kingdom

Technology, Programming, and Applications in Industrial Robotics

ISBN: 978-1-78154-501-0

British Library Cataloguing in Publication Data
A CIP record for this book is available from the British Library

Exclusively distributed by CBS Publishers & Distributors Pvt. Ltd.

Sales & Distribution Rights only for India, Pakistan, Bangladesh, Sri Lanka, Nepal and Bhutan.This book is not to be sold outside these territories.

PREFACE

Robotics brings together several very different engineering areas and skills. There is metalworking for the body. There is mechanics for mounting the wheels on the axles, connecting them to the motors and keeping the body in balance. You need electronics to power the motors and connect the sensors to the controllers. At last you need the software to understand the sensors and drive the robot around.

This book tries to cover all the key areas of robotics as a hobby. When possible examples from industrial robots will be addressed too.

You'll notice very few "exact" values in these texts. Instead, vague terms like "small", "heavy" and "light" will be used. This is because most of the time you'll have a lot of freedom in picking these values, and all robot projects are unique in available materials.

This book is a core-text of industrial engineering undergraduate studies a well referenced, attractively presented, comprehensive users guide to robotics for industrial applications. Beyond that, is the demonstration of the engineering approach to systems development in complex environments (*e.g.* business) that has a far wider application.

This page left intentionally blank.

CONTENTS

Carlos Rodriguez-Donate, Luis Morales-Velazquez, Roque Alfredo Osornio-Rios, Gilberto Herrera-Ruiz and Rene de Jesus Romero-Troncoso

This page left intentionally blank.

List of Contributors

Carlos Rodriguez-Donate

HSPdigital – CA Mecatronica, Facultad de Ingenieria, Universidad Autonoma de Queretaro, Campus San Juan del Rio, Rio Moctezuma 249, 76807 San Juan del Rio, Qro., Mexico; E-Mails: cdonate@hspdigital.org (C.R.-D.); lmorales@hspdigital.org (L.M.-V.); raosornio@hspdigital.org (R.A.O.-R.)

Luis Morales-Velazquez

HSPdigital – CA Mecatronica, Facultad de Ingenieria, Universidad Autonoma de Queretaro, Campus San Juan del Rio, Rio Moctezuma 249, 76807 San Juan del Rio, Qro., Mexico; E-Mails: cdonate@hspdigital.org (C.R.-D.); lmorales@hspdigital.org (L.M.-V.); raosornio@hspdigital.org (R.A.O.-R.)

Roque Alfredo Osornio-Rios

HSPdigital – CA Mecatronica, Facultad de Ingenieria, Universidad Autonoma de Queretaro, Campus San Juan del Rio, Rio Moctezuma 249, 76807 San Juan del Rio, Qro., Mexico; E-Mails: cdonate@hspdigital.org (C.R.-D.); lmorales@hspdigital.org (L.M.-V.); raosornio@hspdigital.org (R.A.O.-R.)

Gilberto Herrera-Ruiz

Facultad de Ingenieria, Universidad Autonoma de Queretaro, Cerro de las Campanas s/n, 76010 Queretaro, Qro., Mexico; E-Mail: gherrera@uaq.mx

Rene de Jesus Romero-Troncoso

HSPdigital – CA Telematica, DICIS, Universidad de Guanajuato, Carr. Salamanca-Valle km 3.5+1.8, Palo Blanco, 36700 Salamanca, Gto., Mexico

This page left intentionally blank.

Chapter 1

FUNDAMENTALS OF ROBOTICS

INTRODUCTION

Robotics brings together several very different engineering areas and skills. There is metalworking for the body. There is mechanics for mounting the wheels on the axles, connecting them to the motors and keeping the body in balance. You need electronics to power the motors and connect the sensors to the controllers. At last you need the software to understand the sensors and drive the robot around.

This book tries to cover all the key areas of robotics as a hobby. When possible examples from industrial robots will be addressed too.

You'll notice very few "exact" values in these texts. Instead, vague terms like "small", "heavy" and "light" will be used. This is because most of the time you'll have a lot of freedom in picking these values, and all robot projects are unique in available materials.

ROBOTICS - ROBOTICS HISTORY

According to the Robot Institute of America (1979) a robot is:

"A reprogrammable, multifunctional manipulator designed to move material, parts, tools, or specialized devices through various programmed motions for the performance of a variety of tasks".

A more inspiring definition can be found in Webster. According to Webster a robot is:

"An automatic device that performs functions normally ascribed to humans or a machine in the form of a human."

First use of the word 'Robot'

The acclaimed Czech playwright Karel Capek (1890-1938) made the first use of the word 'robot', from the Czech word for forced labor or serf. Capek was reportedly several times a candidate for the Nobel prize for his works and very influential and prolific as a writer and playwright.

The use of the word Robot was introduced into his play *R.U.R. (Rossum's Universal Robots)* which opened in Prague in January 1921.

In R.U.R., Capek poses a paradise, where the machines initially bring so many benefits but in the end bring an equal amount of blight in the form of unemployment and social unrest.

The play was an enormous success and productions soon opened throughout Europe and the U.S. R.U.R's theme, in part, was the dehumanization of man in a technological civilization.

You may find it surprising that the robots were not mechanical in nature but were created through chemical means. In fact, in an essay written in 1935, Capek strongly fought that this idea was at all possible and, writing in the third person, said:

"It is with horror, frankly, that he rejects all responsibility for the idea that metal contraptions could ever replace human beings, and that by means of wires they could awaken something like life, love, or rebellion. He would deem this dark prospect to be either an overestimation of machines, or a grave offence against life."

There is some evidence that the word robot was actually coined by Karl's brother Josef, a writer in his own right. In a short letter, Capek writes that he asked Josef what he should call the artificial workers in his new play.

Karel suggests Labori, which he thinks too 'bookish' and his brother mutters "then call them Robots" and turns back to his work, and so from a curt response we have the word robot.

First use of the word 'Robotics'

The word 'robotics' was first used in Runaround, a short story published in 1942, by Isaac Asimov (born Jan. 2, 1920, died Apr. 6, 1992). I, Robot, a collection of several of these stories, was published in 1950.

One of the first robots Asimov wrote about was a robotherapist. A modern counterpart to Asimov's fictional character is Eliza. Eliza was born in 1966 by a Massachusetts Institute of Technology Professor Joseph Weizenbaum who wrote Eliza — a computer program for the study of natural language communication between man and machine.

She was initially programmed with 240 lines of code to simulate a psychotherapist by answering questions with questions.

Three Laws of Robotics

Asimov also proposed his three "Laws of Robotics", and he later added a 'zeroth law'.

Law Zero: A robot may not injure humanity, or, through inaction, allow humanity to come to harm.

Law One: A robot may not injure a human being, or, through inaction, allow a human being to come to harm, unless this would violate a higher order law.

Law Two: A robot must obey orders given it by human beings, except where such orders would conflict with a higher order law.

Law Three: A robot must protect its own existence as long as such protection does not conflict with a higher order law.

The First Robot: 'Unimate'

After the technology explosion during World War II, in 1956, a historic meeting occurs between George C. Devol, a successful inventor and entrepreneur, and engineer Joseph F. Engelberger, over cocktails the two discuss the writings of Isaac Asimov.

Together they made a serious and commercially successful effort to develop a real, working robot. They persuaded Norman Schafler of Condec Corporation in Danbury that they had the basis of a commercial success.

Engelberger started a manufacturing company 'Unimation' which stood for universal automation and so the first commercial company to make robots was formed. Devol wrote the necessary patents. Their first robot nicknamed the 'Unimate'. As a result, Engelberger has been called the 'father of robotics.'

The first Unimate was installed at a General Motors plant to work with heated die-casting machines. In fact most Unimates were sold to extract die castings from die casting machines and to perform spot welding on auto bodies, both tasks being particularly hateful jobs for people.

Both applications were commercially successful, *i.e.*, the robots worked reliably and saved money by replacing people. An industry was spawned and a variety of other tasks were also performed by robots, such as loading and unloading machine tools.

Ultimately Westinghouse acquired Unimation and the entrepreneurs' dream of wealth was achieved. Unimation is still in production today, with robots for sale.

The robot idea was hyped to the skies and became high fashion in the Boardroom. Presidents of large corporations bought them, for about $100,000 each, just to put into laboratories to "see what they could do;" in fact these sales constituted a large part of the robot market. Some companies even reduced their ROI (Return On Investment criteria for investment) for robots to encourage their use.

MODERN INDUSTRIAL ROBOTS

The image of the "electronic brain" as the principal part of the robot was pervasive. Computer scientists were put in charge of robot departments of robot customers and of factories of robot makers. Many of these people knew little about machinery or manufacturing but assumed that they did.

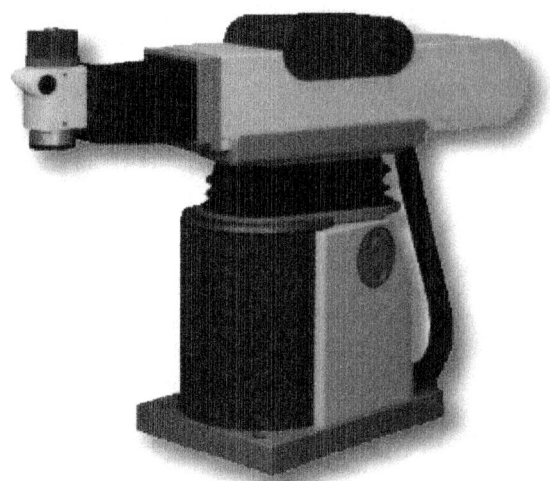

(There is a common delusion of electrical engineers that mechanical phenomena are simple because they are visible. Variable friction, the effects of burrs, minimum and redundant constraints, nonlinearities, variations in work pieces, accommodation to hostile environments and hostile people, *etc.* are like the "Purloined Letter" in Poe's story, right in front of the eye, yet unseen.) They also had little training in the industrial engineer's realm of material handling, manufacturing processes, manufacturing economics and human behavior in factories.

As a result, many of the experimental tasks in those laboratories were made to fit their robot's capabilities but had little to do with the real tasks of the factory.

Modern industrial arms have increased in capability and performance through controller and language development, improved mechanisms, sensing, and drive systems. In the early to mid 80's the robot industry grew very fast primarily due to large investments by the automotive industry.

The quick leap into the factory of the future turned into a plunge when the integration and economic viability of these efforts proved disastrous. The robot industry has only recently recovered to mid-80's revenue levels.

In the meantime there has been an enormous shakeout in the robot industry. In the US, for example, only one US company, Adept, remains in the production industrial robot arm business. Most of the rest went under, consolidated, or were sold to European and Japanese companies.

In the research community the first automata were probably Grey Walter's machina (1940's) and the John's Hopkins beast. Teleoperated or remote controlled devices had been built even earlier with at least the first radio controlled vehicles built by Nikola Tesla in the 1890's.

Tesla is better known as the inventor of the induction motor, AC power transmission, and numerous other electrical devices. Tesla had also envisioned smart mechanisms that were as capable as humans.

An excellent biography of Tesla is Margaret Cheney's Tesla, Man Out of Time, Published by Prentice-Hall, c1981.

SRI's Shakey navigated highly structured indoor environments in the late 60's and Moravec's Stanford Cart was the first to attempt natural outdoor scenes in the late 70's.

From that time there has been a proliferation of work in autonomous driving machines that cruise at highway speeds and navigate outdoor terrains in commercial applications.

Fully functioning androids (robots that look like human beings) are many years away due to the many problems that must be solved. However, real, working, sophisticated robots are in use today and they are revolutionizing the workplace.

These robots do not resemble the romantic android concept of robots. They are industrial manipulators and are really computer controlled "arms and hands". Industrial robots are so different to the popular image that it would be easy for the average person not to recognize one.

Benefits

Robots offer specific benefits to workers, industries and countries. If introduced correctly, industrial robots can improve the quality of life by freeing workers from dirty, boring, dangerous and heavy labor. it is true that robots can cause unemployment by replacing human workers but robots also create jobs: robot technicians, salesmen, engineers, programmers and supervisors.

The benefits of robots to industry include improved management control and productivity and consistently high quality products. Industrial robots can work tirelessly night and day on an assembly line without an loss in performance.

Consequently, they can greatly reduce the costs of manufactured goods. As a result of these industrial benefits, countries that effectively use robots in their industries will have an economic advantage on world market.

WHAT IS ROBOTICS?

A robot is a programmable mechanical device that can perform tasks and interact with its environment, without the aid of human interaction. Robotics is the science and technology behind the design, manufacturing and application of robots.

The word robot was coined by the Czech playwright Karel Capek in 1921. He wrote a play called "Rossum's Universal Robots" that was about a slave class of manufactured human-like servants and their struggle for freedom. The Czech word *robota* loosely means "compulsive servitude." The word robotics was first used by the famous science fiction writer, Isaac Asimov, in 1941.

Basic Components of a Robot

The components of a robot are the body/frame, control system, manipulators, and drivetrain.

Body/frame: The body or frame can be of any shape and size. Essentially, the body/frame provides the structure of the robot. Most people are comfortable with human-sized and shaped robots that they have seen in movies, but the majority of actual robots look nothing like humans. They are typically designed more for function than appearance.

Control System: The control system of a robot is equivalent to the central nervous system of a human. It coordinates and controls all aspects of the robot. Sensors provide feedback based on the robot's surroundings, which is then sent to the Central Processing Unit (CPU). The CPU filters this information through the robot's programming and makes decisions based on logic. The same can be done with a variety of inputs or human commands.

Manipulators: To fulfill their purposes, many robots are required to interact with their environment, and the world around them. Sometimes they are required to move or reorient objects from their environments without direct contact by human operators. Unlike the Body/frame and the Control System, manipulators are not integral to a robot, *i.e.* a robot can exist without a manipulator. This curriculum focuses heavily on manipulators, especially in Unit 6.

Drivetrain: Although some robots are able to perform their tasks from one location, it is often a requirement of robots that they are able to move from location to location. For this task, they require a drivetrain. Drivetrains consist of a powered method of mobility. Humanoid style robots use legs, while most other robots will use some sort of wheeled solution.

Uses and Examples of Robots

Robots have a variety of modern day uses. These uses can be broken down into three major categories:

- Industrial Robots
- Robots in Research
- Robots in Education

Industrial Robots

In industry, there are numerous jobs that require high degrees of speed and precision. For many years humans were responsible for all these jobs. With the advent of robotic technology, it became evident that many industrial processes could be sped up and performed with a higher degree of precision by the use of robots. Such jobs include packaging, assembly, painting, and palletizing. Initially robots in industry only performed very specialized and repetitive jobs that could be executed with a simple yet precise set of instructions. However as technology has improved, we now see industrial robots that are much more flexible, making decisions based on complex sensor feedback. Vision systems are now common on industrial robots. By the end of 2014, the International Federation of Robotics predicts that there will be over 1.3 million industrial robots in operation worldwide!

Robots can be designed to perform tasks that would be difficult, dangerous, or impossible for humans to do. For example, robots are now used to defuse bombs, service nuclear reactors, investigate the depths of the ocean and the far reaches of space.

Robots in Research

Robots come in very handy in the world of research, as they often can be used to perform tasks or reach locations that would be impossible for humans. Some of the most dangerous and challenging environments are found beyond the Earth. For decades, NASA has utilized probes, landers, and rovers with robotic characteristics to study outer space and planets in our solar system.

Pathfinder and Sojourner

The Mars Pathfinder mission developed a unique technology that allowed the delivery of an instrumented lander and a robotic rover, Sojourner, to the surface of Mars. It was the first robotic roving vehicle to be sent to the planet Mars. Sojourner weighs 11.0 kg (24.3 lbs.) on Earth (about 9 lbs. on Mars) and is about the size of a child's wagon. It has six wheels and could move at speeds up to 0.6 meters (1.9 feet) per minute. The mission landed on Mars on July 4th, 1997. Pathfinder not only accomplished this goal but also returned an unprecedented amount of data and outlived its primary design life.

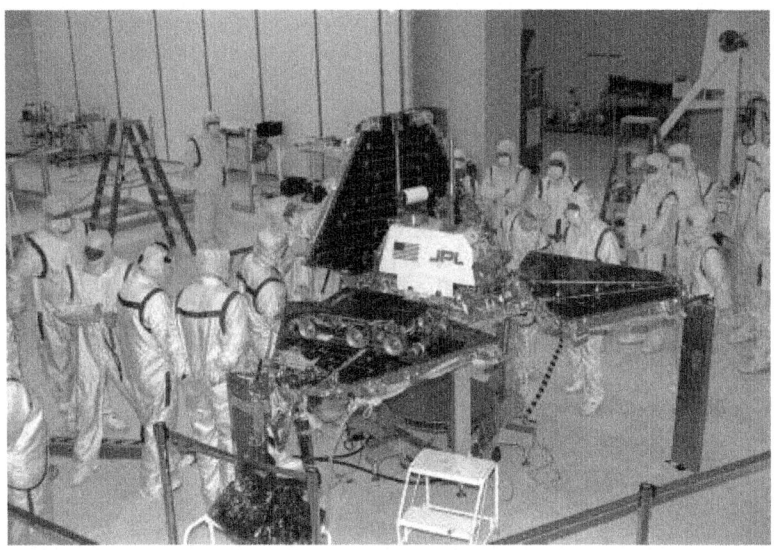

Spirit and Opportunity

The Mars Exploration Rovers (MERs), Spirit and Opportunity, were sent to Mars in the summer 2003 and landed there in January 2004. Their mission was to search for and characterize a wide range of rocks and soils that hold clues to past water activity on Mars in hopes that a manned mission may someday follow. Although initially planned for a lifespan of 90 days, the elapsed mission time surpassed six years, discovering unimaginable amounts of geological information about Mars.

Space Shuttle Robotic Arm

When NASA scientists first began the design for the space shuttle, they realized that there would have to be some way to get the enormous, but fortunately weightless, cargo and equipment into space safely and efficiently. The remote manipulator system (RMS), or Canadarm, made its first flight into space on November 13, 1981.

The arm has six joints, designed to simulate the joints of the human arm. Two are in the shoulder, one is at the elbow, and three are in the highly dextrous wrist. At the end of the wrist is an end effector which can grab or grapple the desired payload. In the weightless environment of space, it can lift more than 586,000 pounds and place it with incredible accuracy. Its total weight on earth is 994 lbs.

The RMS has been used to launch and rescue satellites and has proven itself invaluable in helping astronauts repair the Hubble Space Telescope. The Canadarm's final shuttle mission took place in July of 2011, marking the 90th time it was used on a shuttle mission.

Mobile Servicing System

A similar device to the RMS, the Mobile Servicing System (MSS) otherwise known as Canadarm2 was designed to provide manipulation functions for the International Space Station. The MSS is responsible for servicing payloads and instruments attached to the International Space Station, while also assisting with the transport of supplies and equipment around the station.

Dextre

As part of the Space Shuttle mission STS-123 in 2008, the shuttle Endeavour carried the final part of the Special Purpose Dexterous Manipulator, or "Dextre."

Dextre is a robot with two smaller arms. It is capable of handling the delicate assembly tasks currently performed by astronauts during spacewalks. Dextre can transport objects, use tools, and install and remove equipment on the space station. Dextre also is equipped with lights, video equipment, a tool platform, and four tool holders. Sensors enable the robot to "feel" the objects it is dealing with and automatically react to movements or changes. Four mounted cameras enable the crew to observe what is going on.

Dextre's design somewhat resembles a person. The robot has an upper body that can turn at the waist and shoulders that support arms on either side.

Robots in Education

The field of robotics has become an exciting and accessible tool for teaching and supporting science, technology, engineering, mathematics (STEM), design principles, and problem solving. Robotics enables students to use their hands and minds to create like an engineer, artist, and technician does, all at once. It allows for instantaneous application of scientific and mathematical principals.

In today's education system with its budgetary constraints, middle and high schools are on a constant search for cost-effective exciting ways to deliver high-impact programs that integrate technology with multiple disciplines while preparing students for careers in the twenty-first century. Educators quickly see the advantages that robotics projects and curriculum provide by linking in a cross-curriculum method with other disciplines. Additionally, robotics can provide more affordability and reusability of equipment as compared to other prepackaged options.

Today more than ever, schools are adopting robotics in the classroom to revitalize curriculum and meet ever increasing academic standards required for students. Robotics not only has a unique and broad appeal throughout various teaching fields, but it is quite possibly the technical field that will have the largest influence upon our society throughout the next century.

Why is Robotics Important?

As we saw in the uses and examples of robotics section, robotics is an emerging field with applications in many facets of our lives. It is important for all members of society to have an understanding of the technology that surrounds us. However, robotics is important for more than that reason. Robotics provides a unique combination of the pillars of STEM: science, technology, engineering and math. When taught in schools, it allows students to experience a true interdisciplinary lesson while studying a cutting edge and exciting topic. Also, the aesthetics which go into the design and creation of robots allow students to experiment with an artistic side, while working through technical principals. This combination rewards participants on a plethora of different learning levels.

BASICS OF ROBOT

The vast majority of robots do have several qualities in common. First of all, almost all robots have a movable body. Some only have motorized wheels, and others have dozens of movable segments, typically made of metal or plastic. Like the bones in your body, the individual segments are connected together with joints.

Robots spin wheels and pivot jointed segments with some sort of actuator. Some robots use electric motors and solenoids as actuators; some use a hydraulic system; and some use a pneumatic system (a system driven by compressed gases). Robots may use all these actuator types.

A robot needs a power source to drive these actuators. Most robots either have a battery or they plug into the wall. Hydraulic robots also need a pump to pressurize the hydraulic fluid, and pneumatic robots need an air compressor or compressed air tanks.

The actuators are all wired to an electrical circuit. The circuit powers electrical motors and solenoids directly, and it activates the hydraulic system by manipulating electrical valves. The valves determine the pressurized fluid's path through the machine. To move a hydraulic leg, for example, the robot's controller would open the valve leading from the fluid pump to a piston cylinder attached to that leg. The pressurized fluid would extend the piston, swiveling the leg forward. Typically, in order to move their segments in two directions, robots use pistons that can push both ways.

Fig. : NASA's Urbie climbing stairs.

The robot's computer controls everything attached to the circuit. To move the robot, the computer switches on all the necessary motors and valves. Most robots are reprogrammable — to change the robot's behavior, you simply write a new program to its computer.

Not all robots have sensory systems, and few have the ability to see, hear, smell or taste. The most common robotic sense is the sense of movement — the robot's ability to monitor its own motion. A standard design uses slotted wheels attached to the robot's joints. An LED on one side of the wheel shines a beam of light through the slots to a light sensor on the other side of the wheel. When the robot moves a particular joint, the slotted wheel turns. The slots break the light beam as the wheel spins. The light sensor reads the pattern of the flashing light and transmits the data to the computer. The computer can tell exactly how far the joint has swiveled based on this pattern. This is the same basic system used in computer mice.

These are the basic nuts and bolts of robotics. Roboticists can combine these elements in an infinite number of ways to create robots of unlimited complexity.

The Robotic Arm

The term robot comes from the Czech word *robota*, generally translated as "forced labor." This describes the majority of robots fairly well. Most robots in the world are designed for heavy, repetitive manufacturing work. They handle tasks that are difficult, dangerous or boring to human beings.

The most common manufacturing robot is the robotic arm. A typical robotic arm is made up of seven metal segments, joined by six joints. The computer controls the robot by rotating individual step motors connected to each joint (some larger arms use hydraulics or pneumatics). Unlike ordinary motors, step motors move in exact increments (check out Anaheim Automation to find out how). This allows the computer to move the arm very precisely, repeating exactly the same movement over and over again. The robot uses motion sensors to make sure it moves just the right amount.

An industrial robot with six joints closely resembles a human arm — it has the equivalent of a shoulder, an elbow and a wrist. Typically, the shoulder is mounted to a stationary base structure rather than to a movable body. This type of robot has six degrees of freedom, meaning it can pivot in six different ways. A human arm, by comparison, has seven degrees of freedom.

Your arm's job is to move your hand from place to place. Similarly, the robotic arm's job is to move an end effector from place to place. You can outfit robotic arms with all sorts of end effectors, which are suited to a particular application. One common end effector is a simplified version of the hand, which can grasp and carry different objects. Robotic hands often have built-in pressure sensors that tell the computer how hard the robot is gripping a particular object. This keeps the robot from dropping or breaking whatever it's carrying. Other end effectors include blowtorches, drills and spray painters.

Industrial robots are designed to do exactly the same thing, in a controlled environment, over and over again. For example, a robot might twist the caps onto peanut butter jars coming down an assembly line. To teach a robot how to do its job, the programmer guides the arm through the motions using a handheld controller. The robot stores the exact sequence of movements in its memory, and does it again and again every time a new unit comes down the assembly line.

Most industrial robots work in auto assembly lines, putting cars together. Robots can do a lot of this work more efficiently than human beings because they are so precise. They always drill in the exactly the same place, and they always tighten bolts with the same amount of force, no matter how many hours they've been working. Manufacturing robots are also very important in the computer industry. It takes an incredibly precise hand to put together a tiny microchip.

Mobile Robots

Robotic arms are relatively easy to build and program because they only operate within a confined area. Things get a bit trickier when you send a robot out into the world.

The first obstacle is to give the robot a working locomotion system. If the robot will only need to move over smooth ground, wheels or tracks are the best option. Wheels and tracks can also work on rougher terrain if they are big enough.

But robot designers often look to legsinstead, because they are more adaptable. Building legged robots also helps researchers understand natural locomotion — it's a useful exercise in biological research.

Fig. : Fujitsu's HOAP-1 robot

Typically, hydraulic or pneumatic pistons move robot legs back and forth. The pistons attach to different leg segments just likemuscles attach to different bones. It's a real trick getting all these pistons to work together properly. As a baby, your brain had to figure out exactly the right combination of muscle contractions to walk upright without falling over. Similarly, a robot designer has to figure out the right combination of piston movements involved in walking and program this information into the robot's computer. Many mobile robots have a built-in balance system (a collection of gyroscopes, for example) that tells the computer when it needs to correct its movements.

Fig. : NASA's Frogbot uses springs, linkages and motors to hop from place to place.

Bipedal locomotion (walking on two legs) is inherently unstable, which makes it very difficult to implement in robots. To create more stable robot walkers, designers commonly look to the animal world, specifically insects. Six-legged insects have exceptionally good balance, and they adapt well to a wide variety of terrain.

Some mobile robots are controlled by remote — a human tells them what to do and when to do it. The remote control might communicate with the robot through an attached wire, or using radio or infrared signals. Remote robots, often called puppet robots, are useful for exploring dangerous or inaccessible environments, such as the deep sea or inside a volcano. Some robots are only partially controlled by remote. For example, the operator might direct the robot to go to a certain spot, but not steer it there — the robot would find its own way.

Autonomous Robots

Autonomous robots can act on their own, independent of any controller. The basic idea is to program the robot to respond a certain way to outside stimuli. The very simple bump-and-go robotis a good illustration of how this works.

This sort of robot has a bumper sensor to detect obstacles. When you turn the robot on, it zips along in a straight line. When it finally hits an obstacle, the impact pushes in its bumper sensor. The robot's programming tells it to back up, turn to the right and move forward again, in response to every bump. In this way, the robot changes direction any time it encounters an obstacle.

Advanced robots use more elaborate versions of this same idea. Roboticists create new programs and sensor systems to make robots smarter and more perceptive. Today, robots can effectively navigate a variety of environments.

Simpler mobile robots use infrared or ultrasound sensors to see obstacles. These sensors work the same way as animal echolocation: The robot sends out a sound signal or a beam of infrared light and detects the signal's reflection. The robot locates the distance to obstacles based on how long it takes the signal to bounce back.

Urbie's View

More advanced robots use stereo vision to see the world around them. Two cameras give these robots depth perception, and image-recognition software gives them the ability to locate and classify various objects. Robots might also use microphones and smellsensors to analyze the world around them.

Some autonomous robots can only work in a familiar, constrained environment. Lawn-mowing robots, for example, depend on buried border markers to define the limits of their yard. An office-cleaning robot might need a map of the building in order to maneuver from point to point.

More advanced robots can analyze and adapt to unfamiliar environments, even to areas with rough terrain. These robots may associate certain terrain patterns with certain actions. A rover robot, for example, might construct a map of the land in front of it based on its visual sensors. If the map shows a very bumpy terrain pattern, the robot knows to travel another way. This sort of system is very useful for exploratory robots that operate on other planets (check out JPL Robotics to learn more).

An alternative robot design takes a less structured approach — randomness. When this type of robot gets stuck, it moves its appendages every which way until something works. Force sensors work very closely with the actuators, instead of the computer directing everything based on a program. This is something like an ant trying to get over an obstacle — it doesn't seem to make a decision when it needs to get over an obstacle, it just keeps trying things until it gets over it.

Homebrew Robots

We looked at the most prominent fields in the world of robots — industry robotics and research robotics. Professionals in these fields have made most of the major advancements in robotics over the years, but they aren't the only ones making robots. For decades, a small but passionate band of hobbyists has been creating robots in garages and basements all over the world.

Homebrew robotics is a rapidly expanding subculture with a sizable Web presence. Amateur roboticists cobble together their creations using commercial robot kits, mail order components, toys and even old VCRs.

Homebrew robots are as varied as professional robots. Some weekend roboticists tinker with elaborate walking machines, some design their own service bots and others create competitive robots. The most familiar competitive robots are remote control fighters like you might see on "BattleBots." These machines aren't considered "true robots" because they don't have reprogrammable computer brains. They're basically souped-up remote control cars.

More advanced competitive robots are controlled by computer. Soccer robots, for example, play miniaturized soccer with no human input at all. A standard soccer bot team includes several individual robots that communicate with a central computer. The computer "sees" the entire soccer field with a video camera and picks out its own team members, the opponent's members, the ball and the goal based on their color. The computer processes this information at every second and decides how to direct its own team.

Adaptable and Universal

The personal computer revolution has been marked by extraordinary adaptability. Standardized hardware and programming languages let computer engineers and amateur programmers mold computers to their own particular purposes. Computer components are sort of like art supplies — they have an infinite number of uses.

Most robots to date have been more like kitchen appliances. Roboticists build them from the ground up for a fairly specific purpose. They don't adapt well to radically new applications.

This situation may be changing. A company called Evolution Robotics is pioneering the world of adaptable robotics hardware and software. The company hopes to carve out a niche for itself with easy-to-use "robot developer kits."

The kits come with an open software platform tailored to a range of common robotic functions. For example, roboticists can easily give their creations the ability to follow a target, listen to voice commands and maneuver around obstacles. None of these capabilities are revolutionary from a technology standpoint, but it's unusual that you would find them in one simple package.

The kits also come with common robotics hardware that connects easily with the software. The standard kit comes with infrared sensors, motors, a microphone and a video camera. Roboticists put all these pieces together with a souped-up erector set — a collection of aluminum body pieces and sturdy wheels.

These kits aren't your run-of-the-mill construction sets, of course. At upwards of $700, they're not cheap toys. But they are a big step toward a new sort of robotics. In the near future, creating a new robot to clean your house or take care of your pets while you're away might be as simple as writing a BASIC program to balance your checkbook.

ROBOTS AND ARTIFICIAL INTELLIGENCE

Artificial intelligence (AI) is arguably the most exciting field in robotics. It's certainly the most controversial: Everybody agrees that a robot can work in an assembly line, but there's no consensus on whether a robot can ever be intelligent.

Like the term "robot" itself, artificial intelligence is hard to define. Ultimate AI would be a recreation of the human thought process — a man-made machine with our intellectual abilities. This would include the ability to learn just about anything, the ability to reason, the ability to use language and the ability to formulate original ideas. Roboticists are nowhere near achieving this level of artificial intelligence, but they have made a lot of progress with more limited AI. Today's AI machines can replicate some specific elements of intellectual ability.

Computers can already solve problems in limited realms. The basic idea of AI problem-solving is very simple, though its execution is complicated. First, the AI robot or computer gathers facts about a situation through sensors or human input. The computer compares this information to stored data and decides what the information signifies. The computer runs through various possible actions and predicts which action will be most successful based on the collected information. Of course, the computer can only solve problems it's programmed to solve — it doesn't have any generalized analytical ability. Chess computers are one example of this sort of machine.

Some modern robots also have the ability to learn in a limited capacity. Learning robots recognize if a certain action (moving its legs in a certain way, for instance) achieved a desired result (navigating an obstacle). The robot stores this information and attempts the successful action the next time it encounters the same situation. Again, modern computers can only do this in very limited situations. They can't absorb any sort of information like a human can. Some robots can learn by mimicking human actions. In Japan, roboticists have taught a robot to dance by demonstrating the moves themselves.

Some robots can interact socially. Kismet, a robot at M.I.T's Artificial Intelligence Lab, recognizes human body language and voice inflection and responds appropriately. Kismet's creators are interested in how humans and babies interact, based only on tone of speech and visual cue. This low-level interaction could be the foundation of a human-like learning system.

Kismet and other humanoid robots at the M.I.T. AI Lab operate using an unconventional control structure. Instead of directing every action using a central computer, the robots control lower-level actions with lower-level computers. The program's director, Rodney Brooks, believes this is a more accurate model of human intelligence. We do most things automatically; we don't decide to do them at the highest level of consciousness.

The real challenge of AI is to understand how natural intelligence works. Developing AI isn't like building an artificial heart — scientists don't have a simple, concrete model to work from. We do know that the braincontains billions and billions of neurons, and that we think and learn by establishing electrical connections between different neurons. But we don't know exactly how all of these connections add up to higher reasoning, or even low-level operations. The complex circuitry seems incomprehensible.

Because of this, AI research is largely theoretical. Scientists hypothesize on how and why we learn and think, and they experiment with their ideas using robots. Brooks and his team focus on humanoid robots because they feel that being able to experience the world like a human is essential to developing human-like intelligence. It also makes it easier for people to interact with the robots, which potentially makes it easier for the robot to learn.

Just as physical robotic design is a handy tool for understanding animal and human anatomy, AI research is useful for understanding how natural intelligence works. For some roboticists, this insight is the ultimate goal of designing robots. Others envision a world where we live side by side with intelligent machines and use a variety of lesser robots for manual labor, health care and communication. A number of robotics experts predict that robotic evolution will ultimately turn us into cyborgs — humans integrated with machines. Conceivably, people in the future could load their minds into a sturdy robot and live for thousands of years!

In any case, robots will certainly play a larger role in our daily lives in the future. In the coming decades, robots will gradually move out of the industrial and scientific worlds and into daily life, in the same way that computers spread to the home in the 1980s.

The best way to understand robots is to look at specific designs. The links on the next page will show you a variety of robot projects around the world.

GENERAL DESIGN CONSIDERATIONS

Designing a robot requires balance between size (mostly weight), motor power and battery power. These three elements are connected with each other (more battery power increases the weight of the robot and requires stronger motors) and finding the "perfect" balance requires a lot of tweaking and experimenting. Try to describe heavy components in output/mass (*e.g.* motors: torque/Kg; batteries: mAh/Kg) and pick the one that gives the highest value.

Using light materials brings down the weight significantly (aluminum instead of steel). Building a frame out of light metal and using plastic plates as surfaces would be a lot lighter than using metal plates. For small robots acrylic plastic is a good material to use and it is easy to work with.

There are other ways to build a robot than to cut and drill your own aluminum plates. Toys like Lego Technic and Meccano, although expensive, are an alternative when you don't have the ability to cut and drill your own parts. Especially Meccano (or better: the cheap imitations) can be useful even when you do make your own parts. There is something convenient about having a collection of parts with standard holes and sizes. Of course you would need to drill your own holes at the right distances and size, if you intend to add Meccano parts. The values can (most likely will) differ between different "brands" of imitations. So it's a good idea to buy a few boxes when they are on sale. Screws tend to be M5.

Another kit that would be good to use is the Vex™ Robotic Design Kit from Radio Shack. The parts in this kit are all metal with holes predrilled every half an inch. That makes it easier to add to parts that you may already have. This kit gives you everything you need to get started with robots. If you do not want to buy the kit you could just buy the Vex™ Metal and Hardware Kit for Robotics Starter Kit.

When you start your design, first decide how big you want your robot to be. Don't think about exact sizes, compare it to the size of an object ("the size of a shoe box") will be sufficient. Exact values can be "calculated" after you've got your motors and batteries, as these have a large influence on the size and shape of the robot.

Make an estimate about the weight of the complete robot and pick your motors and wheels. Keep in mind you need high torque and low speed. A bare DC-motor has high speed and low torque, adding a gear reduction will solve this one. Motors with reduction gears are also available. The speed of your motor and the size of your wheels determine how fast your robot will be able to move.

For example: RB-35 is a motor with a 1:50 reduction. It makes 120 RPM or 2 rounds per second. Lets pick a wheel with a diameter of 20cm (radius R = 10cm). This wheel has a circumference of

$$2 \times \pi \times R = 2 \times 3.14 \times 10 = 62.8\text{cm}.$$

This means in one turn the wheel moves 62.8cm. When we mount this wheel on the motor it'll turn twice every second and therefore move

$$2 \times 62.8 \text{ cm} = 125.6\text{cm}.$$

So its speed would be 125.6cm/second or 1.256m/s.

In reality this speed is going to be a little lower as the motor turns 120RPM without a load. But even a 1 m/s is pretty fast for indoor robots. You'll probably use PWM or other methods to slow it down.

Pick your batteries. Make sure you have enough power to keep the motors and all the electronics running for a sufficient amount of time and keep some reserve for future additions. Compare the weight of the batteries and motors you've chosen to what you had planned. You might need to go over this part again (picking different motors and/or batteries).Keep in mind that your robot's body has a significant weight.

PLATFORMS

Wheeled Platforms

Wheeled platforms can have any number of wheels. Most common are 3, 4 and 6 wheeled vehicles (excluding wheels used for feedback). Other numbers are also possible, but can be hard to build, such as 1-wheeled or 2-wheeled robots, or have superfluous wheels which can make turning difficult or complex. Basically there are 2 types of wheels: powered wheels and unpowered wheels. The first are powered by the motors and are used to move the robot forwards (or backwards). Unpowered wheels are used to keep the robot in balance by providing a point of contact with the ground.

Turning

Turning can be accomplished in several different ways:

- Differential Steering (Tank-like Turning):
  - o Moves one wheel forward and the other backwards. The robot turns around within a small circle which center lies in between the 2 powered wheels.
  - o Move one wheel slower than the other, the robot turns in the direction of the slower wheel. How fast it turns depends on how large the difference between the 2 speeds is.
- Ackerman Steering: This is the same steering system as the one used in cars. It is relatively complicated to implement since the inner and outer wheels need to turn to different angles.
- Crab Drive: Each wheel can turn independently in crab drive steering. This can be very flexible, but requires complex mechanics which either turn the entire motor/gearbox/wheel assembly or transfer power from a statically mounted motor. The second option is much more difficult to build but may have advantages over the first.
- 3-wheeled platforms: These can come in a variety of forms, with the articulated wheel powered, or with the two fixed wheels powered, or a combination of the two. These are generally built for very specific purposes.
- Omnidirectional wheels: The omnidirectional wheels design is based upon the use of a series of free turning barrel-shaped rollers, which are mounted in a staggered pattern around the periphery of a larger diameter

main wheel. For this you need 4 powered wheels. However these wheels allow movement in any direction without turning (including sideways and diagonal movement) and can turn the same way as in tanklike steering. Building these wheels is time-consuming, but it's a very powerful steering method. Also, inexpensive omnidirectional wheels are available commercially, often used in conveyors. One drawback, however, is the lack of sideways traction; if something is pushing the robot to the side, it relies on the strength of the motor or brakes to restrain it. Omnidirectional wheels used in place of caster wheels can provide quicker responses and can often roll over larger obstacles.

Tracked Platforms

Tracked platforms use tracks similar to tanks. This kind of propulsion is only useful on loose sand and mud, as concrete and carpet provide too much horizontal traction when turning and will strip the tracks off of their guides.

Walkers

Walkers are robots that use legs instead of wheels or tracks. These robots are harder to build than wheeled robots and can be a nice challenge for an experienced builder. Walkers are designed to imitate how animals (or humans) move.

2-Legged Walkers or Bipeds

This is the hardest type of walker. This type tries to imitate how humans walk. The biggest issue is balance. Two-legged walkers are used for two main purposes: to imitate humans and to provide a great amount of force and traction. Taller walkers used to imitate humans are difficult to build, requiring many balancing circuits and devices, quick motions, and precise construction. Just like any human knows, these can also be knocked over, tripped, *etc.* Shorter, wider walkers can be used to move a large load. When using walkers, it is possible to use pneumatic systems, which can provide a much larger force than motors. However, turning with such a system is nearly impossible.

4-Legged Walkers

4 Legged walkers imitate 4 legged animals. Many of these designs end up moving one leg at a time, instead of the 2-legged movements typical of animals. It requires 3 legs to be on the ground to provide static balance. Dynamic balance moving 2 legs at a time provides faster and more fluid motion.

6-Legged Walkers or Hexapods

These walkers are imitations of insects. Many of these move 3-legs-at-a-time to provide static balance. Because half of the legs can be moved at one time without losing static balance, 6-legged walkers can actually be simpler to build than 4-legged.

Note: Static balance means the construction is at all time in balance. This means that if the robot would stop moving at any time it wouldn't fall over. In contrast there is *Dynamic Balance*. This means that the robot is only in balance when it completes its step. If it's stopped in the middle of its step it would fall over. Although this might sound like a bad thing, dynamic balance allows much faster and smoother movement, but requires sensors to sense balance. Animals and humans move with dynamic balance.

Whegs

There are various combinations of wheels and legs that are useful for varying terrain.

Ball Wheels

This means of propulsion is very similar to how a classic computer mouse works: A ball is mounted in a casing in such a way that it can freely rotate in any direction. Two wheels around the ball are mounted against this ball at an angle of 90° to each other, parallel to the ground. One wheel registers the up-down movements and the other the left-right movements.

A ball wheel uses the same setup but connects the internal wheels to motors. This way the ball can be made to rotate in any direction. A robot equipped with a ball wheel can move up-down and left-right, but can't rotate around its vertical axis. Using 3 ball wheels allows rotation as well.

Electronics

The electronics of a robot generally fall in 6 categories:

- Motor control: controls the movement of the motors, servos and such. Relays and PWM H-bridges fall under this category.
- Sensor reading: reads the sensors and provides this information to the controller.
- Communication: Provides a link between controller and an external PC, another robot, or a remote control.
- Controller: microcontroller board, processor board or logic board. This part makes decisions based on sensor input and the robot's program.
- Power management: parts that provide a fixed 5VDC, 12VDC or any other level coming from the batteries. Circuits that monitor the status of the batteries.
- Glue logic: Additional electronics that allow all the parts to be connected with each other. An example is a CMOS to TTL level converter.

Not all of these categories are present in all robots, nor does every circuit fall completely into one category. Many robots don't require a separate sensor board as a whole lot of sensors have built-in electronics which allow them to connect directly to a μcontroller/processor.

Some Tips

- Use low-power (or simply dimmer) LEDs. Always. This drastically reduces how much current your circuits consume. A normal LED consumes around 15 mA. A modern microcontroller consumes about the same. Disabling unneeded LEDs is advisable, but not always possible.

- Use CMOS ICs instead of classic TTL. Again this reduces current usage and allows a more relaxed supply voltage. But pay attention to their sensitivity for static electricity when soldering them.

- Use good quality IC-sockets (or better don't use IC-sockets at all). They're worth their price.

- Avoid using IC-sockets on sensitive circuits (high speed digital, clock signal and analog signals). The moving robot can shake those ICs loose over time. Using In-Circuit programmable microcontrollers removes the need to be able to unplug an IC.

- LEDs are very practical to make slow digital signals visible, adding them on some signal lines can be interesting for testing purposes, however, they do increase power consumption. Removing them when your circuit works correctly can make it more power efficient (replace LED with a wire, replace resistor with a higher value one).

- See if your microcontrollers can run with a lower clock speed. The higher the clock speed, the more they consume.

- Use your microcontrollers sleep functions whenever possible, disable any part that isn't required (*e.g.* on-chip ADC).

- Learn to make PCBs (Printed Circuit Boards). It's not that hard and it makes your electronics look more professional. Don't throw away your breadboard, PCBs aren't very practical for prototyping.

- If you're up to it: build your circuits in SMD components on PCBs. This reduces size, weight, and cost. However soldering SMD ICs isn't easy. SO-packages (Small Outline) aren't too hard for an experienced builder. Smaller packages are almost impossible to do by hand. Get good soldering tools before you attempt this. For many modern SMD microcontrollers are complete build and tested board available. These can be a solution for those who don't have the equipment, expertise and patience to solder these ICs themselves. These boards don't have to be large as the demo-boards most IC developers sell, *e.g.* the BasicStamp is such a board with the size of an IC.

- Buy a breadboard. They're invaluable for designing and testing circuits.

- If you intent to build your own electronic circuits, invest in a dual channel oscilloscope. Single channel is very restricting. Pick one with as high a bandwidth as you can afford. At least 4x the highest signal frequency you intend to use.
- A good variable power supply is very handy to test how your circuit operates at lower voltage levels (as happens when the batteries discharge).
- If you can choose between pull-up and pull-down resistors, pick the one that uses the lowest amount of power. If the circuits output is at +5V most of the time, use pull-up, if it's 0V use pull-down. Remember that such outputs consume power when the transistor is active (when it's inactive it consumes a small amount of power: a leak current through the resistor and through the input impedance of the next circuit).
- Use high value resistors as pull-up/pull-down. But keep in mind that high speed signal lines need lower value resistor in order to minimize signal distortion.
- Most electronic components on a robot will run on a 5V supply and need a 5V regulator. Use a regulator with low dropout to prevent the 5V supply from browning out.

Display

Very simple robots need only a few LEDs to show everything that it is "thinking".

When trying to debug more complex robot software, it is useful for the robot to display text. A few calculators and PDAs have a RS232 connector or some other simple way to connect to a robot.

Many robots have such a calculator or PDA or other display strapped to the top in order to show the humans what the microcontroller is "thinking", which is vastly more productive than trying to guess what's going on inside that little chip of silicon.

With large robots, sometimes a full-size laptop computer is strapped on top for such display purposes.

Mechanical Design

Balance

Everybody encounters balance every day. While walking, when putting down a glass of water or in so many other ways we have to keep a balance. Now in most of these cases you wouldn't need to think about it, but when designing your robot you'll have to keep an eye on this concept. For most designs, balance isn't hard to achieve, even without doing calculations on it. A few rules of thumb suffice.

If you're into walkers, you'll need to spend more attention to balance. The fewer amount of legs you have, the more important balance becomes.

Simple 4-Wheeled Robots

Achieving balance on this type of robot is pretty trivial. Keep the center of mass between the wheels (picture a rectangle between the centers of the wheels) and as low as possible. In practice it mean you should place the heavy components, *e.g.* the batteries, somewhat in the center of the robot and as low as possible.

Simple 3-Wheeled Robots

These designs are nearly as simple as 4-wheeled robots; the difference lies in that you need to keep the center of mass close to the center of the triangle formed by the wheels. If your robot is rectangular avoid placing weight at the two unsupported corners. These points are prone to making the robot tumble over.

Wheeled Robots with an Arm or Gripper

For the working of an arm or a gripper, we need to take the help of a stepper motor in simple cases or use sensors in sophisticated cases.

DESIGN SOFTWARE

When designing your robot there are plenty of programs to help. Ranging from a simple tool to print wheel encoders, through CAD drawing programs up to mechanical simulation programs.

2D CAD

e.g. AutoCAD. This type of software is used to turn a rough sketch into a nice professional drawing. This type of drawing is standardized for readability. (Meaning every different type of line has a particular meaning. Solid lines are visible edges, dashed lines are hidden edges, line-dash-line lines are center lines. Standards also include methods of dimensioning and types of views presented in a drawing.) Of course you're free to use your own standards, but using an industrial standard, such as ANSI or ISO, makes it easier to share your plans with other people around the world. While it may somewhat more tedious to make a drawing using 2D software, the results are generally better than using 3D solid modeling software. Solid modelers still have problems translating 3D models into 2D drawings and adding proper notation to standards.

- Freeware CAD software: CADstd
- Professional CAD software: AutoCAD
- GPL CAD software: QCad Wikipedia:QCad
- GPL vector drawing software: Inkscape Wikipedia:Inkscape

Solid Modeling

A newer way to draw parts and machines. With solid modeling you "build" the parts in 3D, put them together in an assembly and then let the software generate the 2D drawings (sounds harder than it is). The major advantage over 2D CAD programs is you can see the complete part/machine without actually building it in real life. Mistakes are easily found and corrected in the model. These 3D models are not yet completely standardized though there is a standard for digital data. At this time the 2D drawings this software generates do not conform completely to industrial standards. The 2D paper drawing is still the communication tool of preference in industry and clarity of intent is very important. Solid modeling software tend to generate overly complex drawing views with overly simplified dimensioning methods that likely do not correctly convey the fit, form or function of the part or assembly.

- BRL-CAD c2:BrlCad Wikipedia:BRL-CAD
- lignumCAD (GPL)

Pneumatic & Hydraulic Simulation

Festo has a demo version of both a pneumatic and a hydraulic simulation program. Look for FluidSIM Pneumatiek and FluidSIM Hydraulica.

Limitations: Can't save nor print. Most of the didactic material isn't included.

IRAI has a free demonstration version of electric / pneumatic and hydraulic simulation software : AUTOMGEN / AUTOMSIM. Go to Download / AUTOM-GEN7.

Schematic Capture & PCB

Software to draw electronics schematics and designing Printed Circuit Boards (PCBs). These packages contain software to draw the schematic, libraries with symbols, and software to draw the PCBs (with autorouter).

In no particular order:

- **Freeware:** Eagle is commonly used by beginners for their projects because a limited version is available for free. The toolset is well integrated and has a large hobbiest user base. However, once you progress beyond basic designs, you need to pay for the full version.
- **Open Source:** The open-source gEDA Project has produced a mature suite of applications for electronics design, including: a schematic capture program, attribute manager, netlister supporting over 20 netlist formats, analog and digital simulation, PCB layout with autorouter, and Gerber viewer. The project was started in 1997 to write EDA tools useful for personal robotics projects, but as of this writing the tools are also used by hobbiests, students, educators, and professionals for many different design tasks. The suite runs best on Linux and OSX, although Windows ports of some apps have been made.
- **Open Source:** Kicad, wikipedia: Kicad includes schematic capture and PCB layout

- **Open Source:** Free PCB is a mature Windows only open source PCB drafting tool.
- IntelligentCad.org has a few links to FPGA and PCB design tools (GPL)

Controllers

Programming Languages

There are many different programming languages available for μ Controllers:

1. Assembly: Every μcontroller can be programmed in Assembly. However the differences between μcontrollers can be huge. Assembly gives you the most power of the μcontroller but this power comes with a price: Hard to learn and (almost)no code reuse.

2. Assembly code is in essence translated machine code. It provides only the instruction set of the processor: add, subtract, *maybe* multiply, move data between registers and/or memory, conditional jumps. No loops, complex selection or build in I/O as in C/C++, Basic, Pascal,...

3. The disadvantage is that you have to implement everything yourself (lots of work even for the most simple programs).

4. The advantage is that you have to implement everything yourself (programs can be written extremely efficient both in speed and size).

5. This language is intended for advanced users and is usually only used as an optimisation for code in tight loops or for pushing the performance of a limited device to the edge of its abilities.

6. Reasons to learn it:
   - Teaches you how the computer works on its lowest level.
   - Provides high speed code which consumes little memory.
   - Reasons to avoid it:
   - Limited use.
   - Non-portable.
   - very hard to master.
   - Freeware: AVR

7. C: C offers power but is much more portable than Assembly. For most μ controllers there is a C compiler available. The differences between μ controllers is smaller here, except for using hardware.

8. Learning C is much easier than learning Assembly, still C isn't an easy language to learn from scratch. However these days there are very good books available on this subject.
   - Freeware: GCC Tools for AVR Studio Software
   - Basic: For many μ controllers there are special flavours of Basic available. This is the easiest and fastest way to code μ controllers, however you'll have to sacrifice some power. Still modern basic compilers can produce very impressive code.

- Limited Freeware/payware: Bascom AVR Very good Basic compiler for AVR. Limited to 4Kb programs. There is also a version available for the 8051 µcontrollers.
- Limited Freeware/payware: XCSB PIC Basic compiler. Lite version. No 32-bit integer and floating point support. (OS/2 WARP, Win95, Win98, Win2K, XP and Linux)
- Forth:
- PFAVR (GPL) Needs external RAM.
- ByteForth Dutch and works without external RAM, there is also a building book (Dutch only for now) available for Ushi our robotic project.
- Python
- Pyastra
- PyMite

Programmers

After you've written your program, you need to get it into your µcontroller. If you use C or Basic you'll have to compile it. Then use a programmer to upload the code into the µcontroller. There are several different methods for this last step.

- **External programmers:** This is a device that's connected to a PC. You'll plug the µ controller IC, EEPROM or other memory IC in its socket and let the PC upload the code. Afterwards you plug the IC in its circuit and test it. Can be time consuming when updating your program after debugging.
- **ISP In System Programming:** The board with the µcontroller has a special connector to connect to a PC. Hook up the cable, download code, test and repeat. More modern method. Only disadvantage: it consumes some boardspace. Not all µ controllers support this.
- **Bootloader, also called "self-programming":** The CPU accepts a new program through any available connection to a PC (no special connector needed), then programs itself. Not all µ controllers support this. And you also need some other programming method, to get the initial bootloader programmed in (telling it exactly which connector to watch for a new program, the baud rate, *etc.*).

TOOLS AND EQUIPMENT

Mechanical Tools

For building your robot you'll need some tools to form the body.

1. **Small vise:** you'll need this.
2. **Hammer:** A hammer is one of the standard tools you'll need.

3. **Screwdrivers & Wrenches:** their uses are obvious. Two spanners of equal size are required for locknutting.
4. **Saw:** Metal and wood saws. Miter saws can be very handy, but are pretty expensive. A miter box might suffice for many purposes.
5. Square, measuring tape, scriber and other marking out tools.
6. **Vernier calipers:** Allow very accurate marking out and measurement. Also can be used to check thread pitch on machine screws without a dedicated pitch gauge.
7. **Files:** especially when working with metal, as rough metal edges are sharp.
8. **Centre Punch:** Essential for accurate drilling of holes in metal to prevent the drill skating over the surface.
9. **Drill Press:** (small table top versions suffice) is very handy for drilling accurate holes. Can also provide the low speeds for drilling large holes in metal, which hand drills cannot do easily.
10. **Hobby Tool:** Useful for many purposes.
11. **Sharp utility knifes:** Mostly used when working with plastics.
12. **Hot glue guns:** handy for quickly mounting parts. Not too strong bound, but useful for many applications.
13. **Arc Welder:** Only useful when working with thick steel on large projects (use a gas welding torch for thin metal;arc welders tend to burn holes right through the workpiece). Aluminium cannot be welded with ordinary welders.
14. **Paint stripper/Electric Heat Gun:** like a hairdryer on steroids. Useful for bending plastics, also applying heat-shrink tubing to electric cables at low power.
15. **Safety Goggles:** You only get one pair of eyes, and machine tools are potentially dangerous. Safety goggles are essential for using anything other than hand tools.

ELECTRONIC TOOLS

Soldering Iron

The soldering iron is a very useful tool for assembling electronic circuits and connecting copper wires together.

For electronic circuits you'll need a light soldering iron (~25W) with a small point (shaped like a pencil point). Especially SMD components require small points (or even better: special SMD soldering points).

Soldering electronic components is done with "soft soldering": with a low temperature (less than 300°C). Usually for electronics the melting point of the solder lays around 238°C. When buying solder choose for a solder wire (60% lead, 40% tin) with non-corrosive flux. (There is also "eutectic solder" - 63% lead, 37% tin, which transitions from liquid to solid immediately, with no plastic state in between.) Take the thinnest wire you can find (<=1mm).

For connecting metal wires you'll need something more powerful (30W-100W) like a soldering pistol, but an ordinary soldering iron would do just as well. Note: not all materials are as good to solder. Copper is easy to solder and has a reasonable strong bond. Aluminum has a weak bond.

For stronger connections it's better to braze instead of soft soldering. Brazing involves higher temperatures (typical between 450°C and 1000°C) and different flux ("Borax") and solder (copper and zinc or silver alloys) it also requires a welding torch instead of a soldering iron.

If you need even more strength you could use welding. However welding is only used for heavy materials like steel alloys and these are in most cases too heavy to be used in robots (unless you're building a very big or industrial robot). Aluminum can be welded but it isn't as simple as welding steel alloys.

Breadboard

The boards allow you to build a temporary circuit in no time. Especially handy for testing new circuits. Connections are made with either ordinary thin stiff wire with the insulation removed at the ends or with special breadboard wires with stronger tips. Wires with crocodile clamps are needed for hooking up signal generator, oscilloscope, DMM, *etc.* Larger boards have connectors (typically banana plugs) for the power supply.

There are small breadboards with an adhesive strip at the bottom. These can be mounted on an empty part of a microcontroller board and can be used to build small circuits.

- **Note:** when you build a sensitive analog circuit on a breadboard, it can behave differently than when it's build as a PCB. This is because of parasitic components: the wires connecting the components on the board act as a combination of a resistor, capacitor and coil (all with very low values). Keep in mind that in some circumstances this can affect the working of a circuit. Usually this is only a problem when working with low amplitudes and/or high frequencies.

Electronic Equipment

- **Multimeter:** measures voltage, current & resistance. Many can measure transistor and diode characteristics, frequency and capacity. Some can measure temperature or light intensity.
- **Note:** measuring voltage and current of a AC source isn't as simple as measuring DC levels. But since robots rarely use AC this would be out of the scope of this text. But if you would require to measure AC levels you should read up on this.
- **Oscilloscope:** makes a electric signal visible. Very useful when working with more complicated electronic circuits, especially analog signals and data communication. Oscilloscopes exist as stand-alone devices or as

add-on modules for PCs. The latter provides extra abilities like spectrum analysing and recording of signals.

- **Variable power supplies:** power supplies with variable output. Either AC or DC. Either the output voltage or current can be regulated, although most power supplies let you set a max current.

- **Signal generators:** generates different shapes of signals (sine, square, saw and triangle), with variable frequency (1Hz up to 100MHz) and amplitude.

- **Logic probe:** pen-like devices that detect logic levels (either TTL or CMOS). Most can detect pulse signals. Very handy when working with digital electronic circuits.

- **Frequency meters:** measures the frequency of a signal. Can also be used as a pulse counter. Oscilloscopes can be used for measuring frequency, and storage scopes can freeze a waveform onscreen allowing pulses to be manually counted, but frequency meters are a good investment if this needs to be done very often.

- **LEDs:** An underrated test device for digital circuits. LEDs are far better than voltmeters for digital circuits in some situations, as you can see many input and output values concurrently, without connecting a multitude of voltmeters or constantly checking everything with a logic probe. In particular, they can instantly show the status of several logic signals simultaneously, impossible with a logic probe. Good breadboard building practice also includes an LED for each breadboard to show it is powered up correctly - this can help avoid the potentially frustrating situation of faultfinding a logic circuit that is actually sound, but has an intermittent or noisy power supply. It's also an excellent indicator if a component is short-circuiting at any time during operation, as the LED will likely dim or go out.

CONNECTORS

Insulation Displacement Connectors (IDC)

Assembling parallel ribbon cables from ribbon and the IDC connectors:

Practical tips:

1. Note that IDC ribbon cable is usually not provided with multicoloured or 'rainbow' insulation, but with single-colour insulation — usually grey or white. However it also has a stripe of coloured ink or paint (red or black) down one side, to guide you with connector orientation. If you need to strip away some of the wires of a multi-way cable to suit the IDC connectors you're using, remove them from the side furthest from the ink stripe so it's still present on the cable.

1. It's usual to fit IDC connectors to the cable so their pin 1 end is on the stripe side of the ribbon. This also helps guide you when you're mating

the cable connectors with those on the equipment, knowing that the stripe corresponds with pin 1.

1. Before clamping an IDC connector to a ribbon cable, make sure that the cable grooves are aligned with the contact jaw tips and that they are also aligned with the scallops moulded into the underside of the clamping strip.

2. Make sure too that the connector pin/jaw axis is as close as possible to 90° with respect to the ribbon cable wire axes. If the connector/ribbon angle is not close to 90°, some connections may not be made properly. If the connector is being fitted at the end of a ribbon cable, cutting the end of the ribbon cleanly square first will allow you to use it as a guide.

3. Try to squeeze the IDC connector and its clamping strip together as evenly as possible, so they remain as close as possible to parallel with each other during the operation. This too ensures that all joints are made correctly. The easiest way to squeeze them together evenly is by using a small machine vice or a special compound-action clamping tool.

4. If an IDC connector has a second cable clamping strip, don't attempt to fit this as part of the main assembly. Assemble the main parts of the connector first on the ribbon cable, and only then fit the second clamping strip.

5. When you are bending the ribbon cable around before fitting the second clamping strip, don't pull it hard. This may loosen some of the connections inside the IDC connector. Just bend the ribbon around gently — a small amount of slack won't do any harm, and may in fact protect the IDC connections from strain.

RJ45 Network Connector

These are the connectors used on UTP network cables. A smaller version (RJ11) is used for telephones. You need a crimping tool to attach the connector to the cable. These connectors are very useful for hooking up different PCB with each other. A good use for RJ45 connectors is for making serial (RS232) programming cables for small embedded systems (many credit card terminals use a DB9 to RJ45 cable to download software from a PC during development). If you are building small embedded controller boards an RJ45 can be a handy connector size to use.

ELECTRONIC COMPONENTS

Special Electronic Components

There are many electronic components that aren't described in most electronic textbooks, but are very useful in robotics.

A typical robot needs a heat sink on the power transistors connected to its motors and other actuators, but often does not need a heatsink connected to its CPU.

ANATOMY OF INDUSTRIAL ROBOTS

The RIA (Robotics Industries Association) has officially given the definition for Industrial Robots. According to RIA, "*An Industrial Robot* is a reprogrammable, multifunctional manipulator designed to move materials, parts, tools, or special devices through variable programmed motions for the performance of a variety of tasks."

The *Anatomy* of Industrial Robots deals with the assembling of outer components of a robot such as wrist, arm, and body. Before jumping into Robot Configurations, here are some of the key facts about robot anatomy.

- **End Effectors:** A hand of a robot is considered as end effectors. The grippers and tools are the two significant types of end effectors. The grippers are used to pick and place an object, while the tools are used to carry out operations like spray painting, spot welding, *etc.* on a work piece.
- **Robot Joints:** The joints in an industrial robot are helpful to perform sliding and rotating movements of a component.
- **Manipulator:** The manipulators in a robot are developed by the integration of links and joints. In the body and arm, it is applied for moving the tools in the work volume. It is also used in the wrist to adjust the tools.
- **Kinematics:** It concerns with the assembling of robot links and joints. It is also used to illustrate the robot motions.

The Robots are mostly divided into four major configurations based on their appearances, sizes, *etc.* such as:

- Cylindrical Configuration,
- Polar Configuration,
- Jointed Arm Configuration, and
- Cartesian Co-ordinate Configuration.

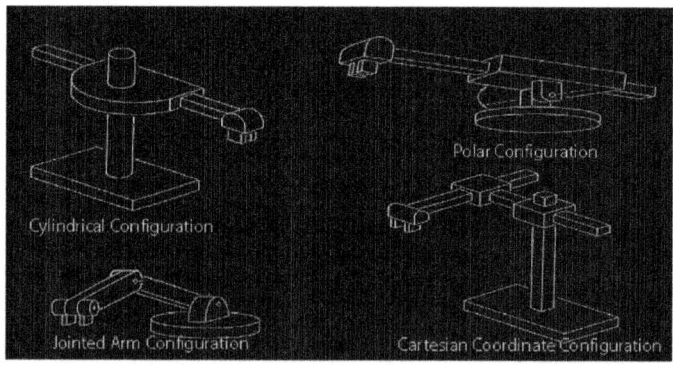

Cylindrical Configuration

This kind of robots incorporates a slide in the horizontal position and a column in the vertical position. It also includes a robot arm at the end of the slide.

Here, the slide is capable of moving in up & down motion with the help of the column. In addition, it can reach the work space in a rotary movement as like a cylinder.

Example: GMF Model M1A Robot.

Advantages:

- Increased rigidity, and
- Capacity of carrying high payloads.

Disadvantages:

- Floor space required is more, and
- Less work volume.

Polar Configuration

The *polar configuration* robots will possess an arm, which can move up and down. It comprises of a rotational base along with a pivot. It has one linear & two rotary joints that allows the robot to operate in a spherical work volume. It is also stated as Spherical Coordinate Robots.

Example: Unimate 2000 Series Robot.

Advantages:

- Long reach capability in the horizontal position.

Disadvantages:

- Vertical reach is low.

Jointed Arm Configuration

The arm in these configuration robots looks almost like a human arm. It gets three rotary joints and three wrist axes, which form into six degrees of freedoms. As a result, it has the capability to be controlled at any adjustments in the work space. These types of robots are used for performing several operations like *spray painting, spot welding, arc welding,* and more.

Example: Cincinnati Milacron T3 776 Robot

Advantages:

- Increased flexibility,
- Huge work volume, and
- Quick operation.

Disadvantages:

- Very expensive,
- Difficult operating procedures, and
- Plenty of components.

Cartesian Co-ordinate configuration

These robots are also called as *XYZ robots*, because it is equipped with three rotary joints for assembling XYZ axes. The robots will process in a rectangular work space by means of this three joints movement. It is capable of carrying high payloads with the help of its rigid structure. It is mainly integrated in some functions like pick and place, material handling, loading and unloading, and so on. Additionally, this configuration adds a name of Gantry Robot.

Example: IBM 7565 Robot.

Advantages:

- Highly accurate & speed,
- Fewer cost,
- Simple operating procedures, and
- High payloads.

Disadvantages:

- Less work envelope, and
- Reduced flexibility.

RETHINKING ROBOT ANATOMY

Actuator-based models offer alternative robot designs.

The snake-like movement of this seven-axis robot allows it to load and unload machine tools while occupying little floorspace. The rendering illustrates a configuration in which the robot could load and unload two machines and still leave the floorspace in front of both of them unrestricted.

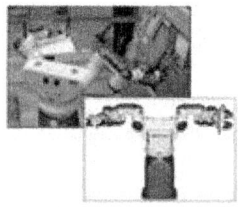

This robot has two independent arms that can perform separate tasks or work together. The two arms working in unison can speed the load/unload cycle, and they can also simplify the gripper design in applications where two workpieces need to be held at once. The upper photo shows the robot simultaneously unloading one part and grabbing the next part to be loaded.

We all know what an industrial robot looks like. It looks vaguely like a human arm. That is, the typical robot used in a machine shop's machine tending application has pivots near the base that work somewhat like a human shoulder, a pivot along the length that resembles an elbow and a pivot near the gripper that resembles a wrist. The design also has two long links in the way that a human has an upper and lower arm. A study of our own limbs, it seems, influenced the design of these machines.

However, a robot doesn't have to look this way. A couple of models from robot supplier Motoman (West Carrollton, Ohio) illustrate alternatives.

The images at left show a seven-axis robot that uses actuator-driven motion. With the seven axes of programmable motion, and with the shorter links, this robot moves more like a snake or tentacle as opposed to a human arm. Standing upright, this robot occupies a footprint just slightly larger than 1 square foot. In fact, it can stand in this footprint close beside a machine tool — or close between two machines — and snake around into the machine's work zone as needed. This configuration leaves the floor space fully open both in front of the machine and behind it.

The use of actuators permits such a compact form. An actuator is a motion-control device that integrates motor, encoder, reducer and brake into a single body. Normally, these would be separate elements of the robot design, and an important advantage of this established design is power. Robots with separate motors can handle very heavy parts. By contrast, the limitation on the actuator-based robot is how much torque an actuator can deliver. The actuator-based robot is therefore flexible, compact and unobtrusive, but at present it is limited to payloads of 20 kg.

Another actuator-based robot is the dual-arm robot shown on the next page. This 13-axis robot has two six-axis arms, and it works within a footprint comparable to the working space of a human being. In fact, this robot arguably does resemble a human being — except that its two arms work independently in ways no human's two arms ever could. Having two independently programmable arms on a robot can dramatically simplify a machine-tending application, because many such applications require complex dual-gripper mechanisms to let a single arm exchange two parts quickly. With two arms working together, simple grippers can accomplish the same quick exchange.

Tom Sipple is the handling products technology leader for Motoman. He says that he and other engineers like him are working to better understand all of the applications in which a dual-arm robot might be applied more effectively than a single arm. Conventional robots are well accepted now, but he predicts that in 10 years, dual-arm robots will also be commonplace. He says one clue that a dual-arm robot may be appropriate in a machining application is when the robot itself is the limiting factor on the cycle time of the overall process. He offers these examples of machine-tending applications where dual-arm robots have proven to be effective:

- Loading a transfer machine. On a high-volume, high-productivity machine such as a dial index machine, the two arms can work simultaneously to unload the finished part while loading the next blank.

- Loading related machines. With the robot between two machine tools (a lathe and a machining center, for example), arm number 1 can hand off the partially machined part to arm number 2. Arm 2 can then load that part in the second machine while arm 1 goes to load the next unmachined part in the first machine.

- Deburring or inspection. While the robot waits for a machining cycle to finish, the two arms can work together to perform a simple operation on a part that has just been machined. Specifically, one arm can hold the part while the other uses a deburring tool or a gage.

- Quick part exchange in small machines. The work zones of some machine tools are so restricted that they don't even offer clearance for dual-gripper devices on a single arm. For machines such as these, dual arms introduce a way to unload one part and load the next part in a significantly shorter span of time.

ANATOMY OF INDUSTRIAL ROBOTS

There are several classes of robots: robotic aircraft, robotic ships, mobile robots and others. An important application of robots is in industry – for machine tending, welding, painting, assembly and *etc.* These "industrial robots" can be viewed as consisting of a mechanical portion "the manipulator" controlled by a microprocessor.

Subsystems of Industrial Robots Include

Actuators

Actuators are basically prime movers providing both force and motion. Pneumatic cylinders, hydraulics, permanent magnet motors, stepper motors, linear motors are some conventional actuators. More advanced ones are based on hi-tech polymers, shape memory alloys, piezo patches, and pneumatic muscles. Brushless servo motors also exist for low noise levels, and printed armature motors are used for quick response.

Transmission Systems

The transmission system used in robot to transmit power and motion consists of chains, timing belts, metal belts, cables and pulleys and linkages. Gear boxes and harmonic drives serve to provide speed reduction. Ball screws are used with suitable mechanisms to convert rotary motion to linear motion and if needed back to oscillatory motion. Drive stiffness is an important consideration in robotics and so also is backlash.

Power Supplies

Hydraulic and Pneumatic power packs: These consist of a motor driving a positive displacement pump or compressor to generate the high pressure fluid flow. In using hydraulic systems the necessity of having an oil tank increases the weight of the system, additionally the issue of ensuring that the oil is free of contaminants is to be handled. In pneumatics power pack dry air is desired.

Electric motors use what ate known as PWM (pulse width modulation) amplifiers. These are electronic devices, consisting of transistors used as switches to rapidly switch on and off the supply in a controlled manner to control motor speeds. Such drives have higher efficiency.

Sensors and Other Electronics

The sensors for feedback in robots consists of tachometers and encoders and potentiometers to sense motor motions, simple switches, force sensors, acceleration sensors, optical systems, special cameras and vision systems.

Electronics

There are a host of electronic circuits, motor controllers, analog to digital converters and digital to analogue converters, frame grabbers and so on utilized to handle sensors and vision systems and convert the inputs from them into a form usable by the processor for control of the entire system in conjunction with the algorithms and software developed specifically for the purpose.

Software

The software used consists of several levels. Motor control software consists of algorithms which help the servo to move smoothly utilizing the data from feed-back units. At the next level there is software to plan the trajectory of the end effector and translate the same into commands to individual motor controllers. The output of sensors is also to be interpreted and decisions made. At the highest level there is software which accepts commands from the user of the robot and translates it into appropriate actions at the lower level.

ROBOTICS - CURRENT RESEARCH

The robots of tomorrow will be the direct result of the robotic research projects of today. The goals of most robotic research projects is the advancement of abilities in one or more of the following technological areas:

Artificial intelligence, effectors and mobility, sensor detection and especially robotic vision, and control systems.

These technological advances will lead to improvements and innovations in the application of robotics to industry, medicine, the military, space exploration, underwater exploration, and personal service.

ARTIFICIAL INTELLIGENCE

Human Behavior and Emotion

Fig. : Cog, a humanoid robot from MIT.

Two of the many research projects of the MIT Artificial Intelligence department include an artificial humanoid called Cog, and his baby brother, Kismet. What the researchers learn while putting the robots together will be shared to speed up development.

Once finished, Cog will have everything except legs, whereas Kismet has only a 3 6-kilogram head that can display a wide variety of emotions. To do this Kismet has been given movable facial features that can express basic emotional states that resemble those of a human infant. Kismet can thus let its "parents" know whether it needs more or less stimulation — an interactive process that the researchers hope will produce an intelligent robot that has some basic "understanding" of the world.

This approach of creating AI by building on basic behaviors through interactive learning contrasts with older methods, in which a computer is loaded with lots of facts about the world in the hope that intelligence will eventually emerge.

Cog is 2 meters tall, complete with arms, hands and all three senses — including touch-sensitive skin. Its makers will eventually try to use the same sort of social interaction as Kismet to help Cog develop intelligence equivalent to that of a two-year-old child.

Kismet is an autonomous robot designed for social interactions with humans and is part of the larger Cog Project. This project focuses not on robot-robot interactions, but rather on the construction of robots that engage in meaningful social exchanges with humans. By doing so, it is possible to have a socially sophisticated human assist the robot in acquiring more sophisticated communication skills and helping it learn the meaning these acts have for others.

Fig. : Kismet with its creator, Cybthia Breazeal of MIT. Breazeal also helped create Cog.

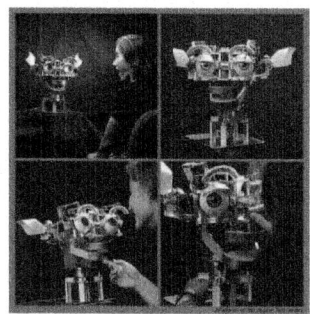

Kismet has a repertoire of responses driven by emotive and behavioral systems. The hope is that Kismet will be able to build upon these basic responses after it is switched on or "born", learn all about the world and become intelligent.

Crucial to its drives are the behaviors that Kismet uses to keep its emotional balance. For example, when there are no visual cues to stimulate it, such as a face or toy, it will become increasingly sad and lonely and look for people to play with.

Any advances made with Kismet will be passed on to its big brother Cog, the robot brainchild of Rodney Brooks, head of MIT's AI department.

Hardware and Software Brains

In mimicking human intelligence, the goal is to make sure robots get a brain and reasoning. An important pioneer in the field of AI is Marvin Minsky.

Without a brain capable of processing input, a robot cannot react to its environment. A brain can be stimulated in hardware or software. Most robots at present have software brains, meaning a computer with a program running. These robots are connected to or equipped with a computer. A drawback is the limited number of processes that can be run on today's computers and the single purpose programs running on these computers. The programs cannot change themselves. In other words, learning is not possible.

A brain made out of hardware, or a number of processors will be closer to reality. The brain will consists of several chips that act both independently and as a group. The general belief is that the real brain works as a neural network of lots of independent processing units. Every chip in itself has a small program. It will process information but also pass it on to other chips. The program changes on a continuous basis. The network of chips is quick and will adapt, so in contrast with the software brain, it will learn.

An example of a hardware brain is Robokoneko the robocat from Genobyte. It has a brain from a machine, the CAM-machine.

EFFECTORS AND MOBILITY

Autonomous Flying Vehicle Project

Robot helicopter research began at the University of Southern California in 1991 with the formation of the Autonomous Flying Vehicle Project and continues to the present day. The first robot built was the AFV (Autonomous Flying Vehicle). The AVATAR (Autonomous Vehicle Aerial Tracking And Retrieval), was created in 1994. The current robot, the second generation AVATAR (Autonomous Vehicle Aerial Tracking And Reconnaissance), was developed in 1997. The "R" in AVATAR changed to reflect a change in robot capabilities.

Fish Robot

Without question, the fish is the best swimmer in the world. That is why the Ship Research Institute of Japan decided to build the Fish Robot. This project hopes to apply what is learned while building and researching with the Fish Robot to the design and construction of ships.

Fish Robot PF-600 with Tuna-type Tail Fin

Muscles

Robots use electro-engines for movement. Engine parts are relatively cheap and last long. Engines are applied to move arm, turn wheels or move other parts, for instance camera's. Engines are less usefull with walking robots. In that particular case engines prove to be a weak part, a jumping robot is a mayor challenge to engine parts. Human being use muscle, which contract and expand, to move around. A muscle receive a signal form the brain and contracts. Causing a joint, like the knee to move.

Material to mimic a muscle is still a dream. Nitinol, an alloy that consist of the metals nickel and titanium will shrink if an electric current travels through the alloy, it will only contract 8% maximum.

The downside, nitinol is very expensive en the contraction is too little to allow it to be used to make walking robots. For the time being walking robots will not use muscles or engines but pneumatic of hydrolic technologies.

Robocup

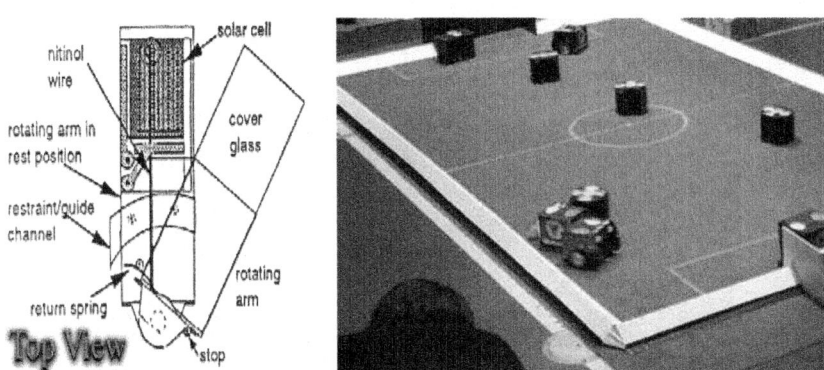

To demonstrate advances in research and to stimulate scientist to share progress the Robocup competition is organized a few times a year. Robocup is a competition of Robot soccer teams. Movement, pattern recognition, where's the ball, where's the goal, who is in my team, all this and more is needed to score a goal. A simple games becomes a challenge for a robot team. Besides moving and finding the ball and team members the robots needs to define a strategy and take lots of decisions in a short time frame. Robocup has produced many advancements in both robotic effectors and sensors. Who could have imagined that soccer would contribute to robot research where robots eventually will be smart and capable of cooperation with other to reach a goal?

SENSOR DETECTION

Robotic Vision

Machine vision involves devices which sense images and processes which interpret the images. Thomas Braunl of The University of Western Australia provides an excellent example of robotic vision research in both hardware and software.

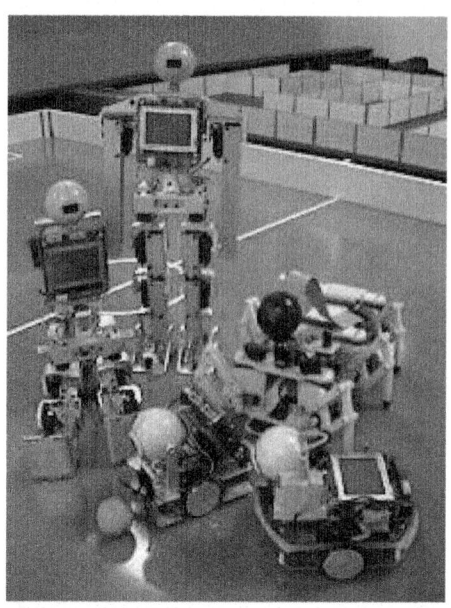

EyeBot is a controller for mobile robots with wheels, walking robots or flying robots. It consists of a powerful 32-Bit microcontroller board with a graphics display and a digital grayscale or color camera. The camera is directly connected to the robot board (no frame grabber). This allows programmers to write powerful robot control programs without a big and heavy computer system and without having to sacrifice vision.

- Ideal basis for programming of real time image processing
- Integrated digital camera (grayscale or color)
- Large graphics display (LCD)
- Can be extended with own mechanics and sensors to full mobile robot
- Programmed from IBM-PC or Unix workstation,

- programs are downloaded *via* serial line (RS-232) into RAM or Flash-ROM
- Programming in C or assembly language

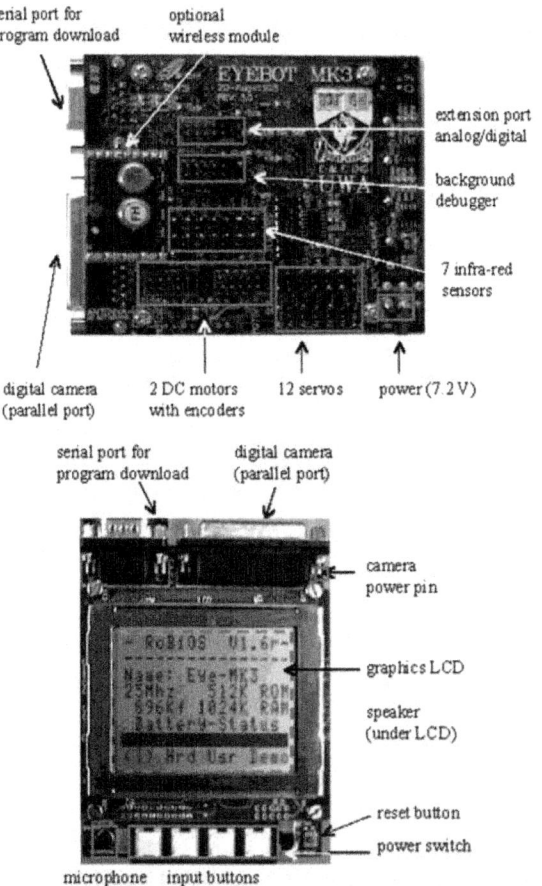

serial port for program download
optional wireless module
extension port analog/digital
background debugger
7 infra-red sensors
digital camera (parallel port)
2 DC motors with encoders
12 servos
power (7.2 V)

serial port for program download
digital camera (parallel port)
camera power pin
graphics LCD
speaker (under LCD)
reset button
power switch
microphone
input buttons

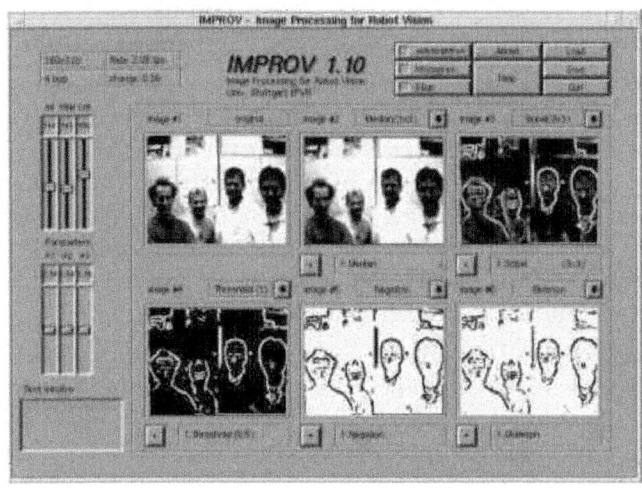

Improv is a tool for basic real time image processing with low resolution, *e.g.* suitable for mobile robots. It has been developed for PCs with *Linux* operating system. *Improv* works with a number of inexpensive low-resolution digital cameras (no framegrabber required), and is available from Joker Robotics. *Improv* displays the live camera image in the first window, while subsequent image operations can be applied to this image in five more windows. For each sub-window, a sequence of image processing routines may be specified.

Sensor Based Motion Planning

Sensor Based Planning incorporates sensor information, reflecting the current state of the environment, into a robot's planning process, as opposed to classical planning, where full knowledge of the world's geometry is assumed to be known prior to the planning event. Sensor based planning is important because:

(1) the robot often has no a priori knowledge of the world;

(2) the robot may have only a coarse knowledge of the world because of limited memory;

(3) the world model is bound to contain inaccuracies which can be overcome with sensor based planning strategies; and

(4) the world is subject to unexpected occurrences or rapidly changing situations.

There already exists a large number of classical path planning methods. However, many of these techniques are not amenable to sensor based interpretation. It is not possible to simply add a step to acquire sensory information, and then construct a plan from the acquired model using a classical technique, since the robot needs a path planning strategy in the first place to acquire the world model.

The first principal problem in sensor based motion planning is the *find-goal* problem. In this problem, the robot seeks to use its on-board sensors to find a collision free path from its current configuration to a goal configuration. In the first variation of the find goal problem, which we term the *absolute find-goal*problem, the absolute coordinates of the goal configuration are assumed to be known.

The second principal problem in sensor based motion planning is *sensor-based exploration*, in which a robot is not directed to seek a particular goal in an unknown environment, but is instead directed to explore the apriori unknown environment in such a way as to see all potentially important features. The exploration problem can be motivated by the following application. Imagine that a robot is to explore the interior of a collapsed building, which has crumbled due to an earthquake, in order to search for human survivors. It is clearly impossible to have knowledge of the building's interior geometry prior to the exploration.

Thus, the robot must be able to see, with its on-board sensors, all points in the building's interior while following its exploration path. In this way, no potential survivors will be missed by the exploring robot. Algorithms that solve the find-goal problem are not useful for exploration because the location of the "goal" (a human survivor in our example) is not known. A second variation on the find-goal problem that is motivated by this scenario and which is an intermediary between the find-goal and exploration problems is the *recognizable* find-goal problem.

In this case, the absolute coordinates of the goal are not known, but it is assumed that the robot can recognize the goal if it becomes with in line of sight. The aim of the recognizable find-goal problem is to explore an unknown environment so as to find a recognizable goal. If the goal is reached before the entire environment is searched, then the search procedure is terminated.

CONTROL SYSTEMS

Hierarchical Behavior Control

Development of a hierarchical behavior control scheme for rovers and mobile robots is currently underway. It attempts to model and control mobile systems using distinct rule-based controllers and decision-making subsystems that collectively represent a hierarchical decomposition of autonomous vehicle behavior. This research approach employs fuzzy logic, behavior control, and genetic programming as tools for developing autonomous robots. Complex, multi-variable fuzzy rule-based systems are developed in the framework of behavior-based control for autonomous navigation. Genetic programming methods are used to computationally evolve fuzzy coordination rules for low-level motion behaviors. In addition, embedded control applications are being developed for microrover navigation using conventional microprocessors and specialized fuzzy VLSI chips.

Nano Technology and Medical Applications

The movie Innerspace shows a miniature spaceship travelling through the artery system of a human. It is a nice illustration of the promise of Nano technology. Nano Technology is a technique where miniature robots go to places humans will never be able to travel. Nano technology is a new science where robotics play a mayor part.

Questions that needs to be solved because of the very tiny mechinical parts: can a robot repair itself, how do you control a nano robot, how does a nano robot move. Will it be able to work autonomously. Will it be able so shift in shape. Is a nanan robot a mechanical device or is it more like a microprocessor. Once these questions are answered Nano technology will change medical science for ever. Surgery will be performed in lots of cases by one or more Nano robots that will travel inside the human body.

Intelligent Systems for Communication Networks

Third generation wireless networks like the Universal Mobile telecommunication System (UMTS) are being developed to support wide band services. A major scenario is to support such services of a user roaming between a cellular terrestrial network and a satellite Personal Communication Network (PCN) while maintaining the quality of service during the hand-over process. And requiring some degree of continuity of quality of service guarantees. This project will focus on developing new protocols which uses artificial intelligent systems to support such hand-over process. Further Seamless roaming and user tracking using intelligent systems will be investigated.

Active Vibration Control

In recent years, the reduction of undesirable vibrations in the dynamic systems such as airplanes, vehicles, tall buildings and off-shore structures has become a crucial issue due to the increased social awareness of comfort as well as the ever increasing heights of new inner city buildings. With the advent of new construction materials and new construction methods, the buildings and structures are becoming taller, and more flexible. With a good design and under normal loading conditions, the response of these structures to vibrations will remain in the safe and comfortable domain.

However, there is no guarantee that in-service loads experienced by tall buildings and structures will always be in the allowed range. The undesirable vibration levels could be reached under large environmental loads such as winds and earthquakes, and could adversely affect human comfort and even structural safety. It is becoming critically important to suppress dynamic responses of tall buildings and structures due to the strong winds and earthquakes not only for their safety but also their serviceability.

When tall buildings and structures are flexible, design performances may become impossible to achieve by conventional design practice. Hence, additional devices are installed in tall buildings and structures to compensate the dynamic responses caused by environmental loads. As a result, new concepts and methods of structural protection have been proposed.

Due to recent development of sensors and digital control techniques, active control methods of dynamic responses of tall buildings and structures have been developed, and some of them have been implemented to actual buildings. The precondition is however that the implementation is simple enough to be realtime. In engineering applications with rule based systems providing efficient results, the implementation is often easier than its complex conventional counterpart.

The active vibration control in the structural engineering has become known as an area of research in which the vibrations and motions of the tall buildings and structures can be controlled or modified by means of the actions of a control system through some external energy supply. Compared with the passive vibration control the active vibration control can more effectively keep the tall buildings and structures safe and comfortable under the various environmental loads such as strong winds or earthquake hazards.

This implies that the active vibration control can be effective and adaptive over a much wider frequency range and also for transient vibration, which is the reason to attract interest of the researchers not only in structural engineering but also in control engineering. Among many methods, that have been proposed, are active mass drivers (AMDs), active tendon systems (ATS), and active variable stiffness systems (AVSs).

Hyper-Redundant Robotics Systems

Robot manipulators which have more than the minimum number of degrees-of-freedom are termed "kinematically redundant," or simply "redundant." Redundancy in manipulator design has been recognized as a means to improve manipulator performance in complex and unstructured environments. "Hyper-redundant" robots have a very large degree of kinematic redundancy, and are analogous in morphology and operation to snakes, elephant trunks, and tentacles. There are a number of very important applications where such robots would be advantageous.

While "snake-like" robots have been investigated for nearly 25 years, they have remained a laboratory curiousity. There are a number of reasons for this:

(1) previous kinematic modeling techniques have not been particularly efficient or well suited to the needs of hyper-redundant robot task modeling;

(2) the mechanical design and implementation of hyper-redundant robots has been perceived as unnecessarily complex; and

(3) hyper-redundant robots are not anthropomorphic, and therefore pose interesting programming problems.

Our research group has undertaken a broadly based program to overcome the obstacles to practical deployment of hyper-redundant robots.

ROBOTICS - ROBOTICS PLANNING

Definitions

Robot

A versatile mechanical device equipped with actuators and sensors under the control of a computing system. Russell and Norvig define it as "an active, artificial agent whose environment is the physical world."

Task Planner

A program that converts task-level specifications into manipulator-level specifications. Thetask planner must have a description of the objects being manipulated, the task environment, the robot, and the initial and desired final states of the environment. The output should be a robot program that converts the initial state into the desired final state.

Guarded Motion

A motion of the form **move along** *path* **until** *sensory-condition*.

Compliant Motion

A motion of the form **move along direction** d **with force = 0 perpendicularly to** d. More intuitively, moving along a surface while maintaining a fixed pressure, such as when scraping paint off of a window with a razor.

Effector

The bits the robot does stuff with. That is, arms, legs, hands, feet. An *end-effector* is a functional device attached to the end of a robot arm (*e.g.*, grippers). actuator.

A device that converts software commands into physical motion, typically electric motors or hydraulic or pneumatic cylinders.degree of freedom.

A dimension along which the robot can move itself or some part of itself. Free objects in 3-space have 6 degrees of freedom, three for position and three for orientation.sensors.

Devices that monitor the environment. There are contact sensors (touch and force), and non-contact (*e.g.*, sonar).

Sonar

Sensing system that works by measuring the time of flight of a sound pulse to be generated, reach an object, and be reflected back to the sensor. Wide angle but reasonably accurate in depth (the wide angle is the disadvantage).infrared

Very accurate angular resolution system but terrible in depth measurement. recognizable set.

An envelope of possible configurations a robot may be in at present; recognizable sets are to continuous domains what multiple state sets are to discrete ones. A recognizable set with respect to a sensor reading is the set of all world states in which the robot might be upon receiving that sensor reading.landmark

An easily recognizable, unique element of the environment that the robot can use to get its bearings.

Motivations

The important incentives for building robots are social, replacing humans in undesirable or dangerous jobs, and economic, reducing the cost of manufacturing while improving its quality.

The real world has the following qualities, that any robot design must take into account:

- **inaccessible** — sensors are imperfect, and can only perceive local stimuli
- **nondeterministic** — the robot can never be certain an action will work as expected, since wheels slip, batteries run down, *etc.*

- **nonepisodic** — the effects of an action change over time, so robots should handle sequential decision problems and learning
- **dynamic** — a robot has to know when to think and when to act right away
- **continuous** — states and actions are drawn from a continuum of physical configurations and motions

In general, robots should have the following qualities:

- **high reliability** — if a robot fails, it should be able to recover or to call for help
- **high speed** — a robot should perform its functions as quickly as needed
- **programmability** — the robot should be flexible and easily adaptable to various tasks
- low cost

The ultimate goal is to build *autonomous* robots that accept commands telling them *what* to do, without needing to specify exactly *how*. Task Planning

By virtue of their versatility, robots can be difficult to program, especially for tasks requiring complex motions involving sensory feedback. In order to simplify programming, *task-level* languages exist that specify actions in terms of their effects on objects.

Example: pin A programmer should be able to specify that the robot should put a pin in a hole, without telling it what sequence of operators to use, or having to think about its sensory or motor operators.

Task planning is divided into three phases: modeling, task specification, and manipulator program synthesis.

There are three approaches to specifying the model state:

1. Using a CAD system to draw the positions of the objects in the desired configuration.
2. Using the robot itself to specify its configurations and to locate the object features.
3. Using symbolic spatial relationships between object features (such as (face1 against face2). This is the most common method, but must be converted into numerical form to be used.

One problem is that these configurations may *overconstrain* the state. Symmetry is an example; it does not matter what the orientation of a peg in a hole is. The final state may also not completely specify the operation; for example, it may not say how hard to tighten a bolt.

The three basic kinds of motions are free motion, guarded motion, and compliant motion.

An important part of robot program synthesis should be the inclusion of sensor tests for error detection. Motion Planning

The fundamental problem in robotics is deciding what motions the robot should perform in order to achieve a goal arrangement of physical objects. This turns out to be an extremely hard problem.

MOTION PLANNING DEFINITIONS

Basic Motion Planning Problem

Let **A** be a single rigid object (the robot) moving in a Euclidean space **W**, called the *workspace*, represented as $\mathbf{R}^n$ (where n = 2 or 3). Let $\mathbf{B}_1,...,\mathbf{B}_q$ be fixed rigid objects distributed in **W**. These are called *obstacles*.

Assume that the geometry of **A**, and the geometries and locations of the $\mathbf{B}_i$'s are accurately known. Assume also that no kinematic constraints limit the motions of **A** (so that **A** is a *free-flying object*).

Given an initial position and orientation and a goal position and orientation of **A** in **W**, the problem is to generate a path **t** specifying a continuous sequence of positions and orientations of **A** avoiding contact with the $\mathbf{B}_i$'s.

(Basically, given a robot, a bunch of objects, a start state and a goal state, find a path for the robot to reach the goal state.) configuration of object A

A specification of the position of every point in the object, relative to a fixed frame of reference. To specify the configuration of a rigid object **A**, it is enough to specify the position and orientation of the frame $\mathbf{F}_A$ with respect to $\mathbf{F}_W$. The subset of **W** occupied by **A** at configuration q is denoted by $\mathbf{A}(q)$. configuration space of object A

The space **C** of all configurations of **A**. The idea is to represent the robot as a point and thus reduce the motion planning problem to planning for a point. dimension of C

The dimension of a configuration space is the number of independent parameters required to represent it as $\mathbf{R}^m$. This is 3 for 2-D, and 6 for 3-D.chart

A representation of a local portion of the configuration space. **C** can be decomposed into a finite union of slightly overlapping *patches* called *charts*, each represented as a copy of $\mathbf{R}^m$.Distance between configurations

The distance between configurations q and q' should decrease and tend to zero when the regions $\mathbf{A}(q)$ and $\mathbf{A}(q')$ get closer and tend to *coincide*.Path

A path from a configuration q_{init} to configuration q_{goal} is a continuous map $t : [0,1] \to \mathbf{C}$ with $t(0) = q_{init}$ and $t(1) = q_{goal}$.free-flying object

An object for which, in the absence of any obstacles, any path is feasible.C-obstacle

An obstacle mapped into configuration space. Every obstacle $\mathbf{B}_i$ is mapped to the following region in the workspace called the C-obstacle: $\mathbf{CB}_i = \{ q$ in $\mathbf{C} : \mathbf{A}(q)$ intersected with $\mathbf{B}_i \mathrel{!=}$ empty set $\}$. C-obstacle region

The union of all the C-obstacles.Free space

All of the configuration space less the C-obstacle region, called C_{free}. A configuration in C_{free} is called a *free configuration* and a *free path* is a path where t maps to C_{free} instead of to **C**. A *semi-free path* maps to the closure of C_{free}.kinematics

Motions in the abstract, without reference to force or mass. Russell and Norvig define it as the study of the correspondence between the actuator motions in a mechanism and the resulting motion of its parts.dynamics

Motions of material bodies under the action of forces.force-compliant motion

Motions in which the robot may touch obstacle surfaces and slide along them. These are more accurate than position-controlled commands.Rotary motion

Rotation around a fixed hub.prismatic motion

Linear movement, as with a piston in a cylinder.

Skeletonization (Roadmap) Methods

Skeletonization methods collapse the configuration space into a one-dimensional subset, or*skeleton*. They then require that paths lie along the skeleton. If the initial and goal points do not lie on the skeleton, short connecting paths are added to join them to the nearest points on the skeleton.

To be complete for motion planning, skeletonization methods must satisfy two properties:

1. If **S** is a skeleton of free space **F**, then **S** should have a single connected piece within each connected region of **F**.
2. For any point p in **F**, it should be "easy" to compute a path from p to the skeleton.

Roadmaps are a *global* approach to motion planning. It consists of modeling the connectivity of therobot's free space as a network of one-dimensional curves, called the *roadmap*, which lies in thefree space or its closure. (**NOTE** that weird things happen at the boundaries depending on which option you choose.)

Once a roadmap **R** has been generated, it is used as a set of standardized paths. Path planning is then reduced to connecting the initial and goal configurations to points in **R**, and searching **R** for apath to connect those points.

The kinds of skeletonizations are *visibility graphs*, *Voronoi diagrams* and *roadmaps*, which consist of *silhouette curves* and *linking curves*. (This formulation due to Russell and Norvig; note that Latombe calls all of these things "roadmap methods.")

Visibility Graphs

The original roadmap method, visibility graphs apply to 2D configuration spaces with polygonal C-obstacle regions. The graph is nondirected and the nodes are the initial and goal configurations plus all the vertices of the C-obstacles. The edges are all the straight line segments connecting two nodes that do not intersect the interior of the C-obstacle region (all lines that don't go through obstacles). The graph can be searched for the shortest semi-free path; this path will always be polygonal. Visibility graphs are complete.

Voronoi Diagram

A roadmap method based on *retraction*. A continuous function of C_{free} is defined onto a one-dimensional subset of itself. The Voronoi diagram consists of those curves in the space that are equidistant from two or more obstacles; these curves form the skeleton.

Put poorly in another way, he Voronoi diagram is the set of all free configurations whose minimal distance to the C-obstacle region is achieved with at least two points in the boundary of the region. This method yields free paths which tend to maximize the clearance between the robot and the obstacles.

When the C-obstacles are polygons, the Voronoi diagram consists of straight and *parabolic segments*. The initial and goal configurations are mapped to q_{init} and q_{goal} by drawing the line along which the distance to the boundary of the C-obstacles increases the fastest.

Cell Decomposition

A global approach to motion planning. The intuitive idea is to break the space down into a finite number of discrete chunks.

Cell decomposition breaks free space down into simple regions called *cells* such that any pathbetween any two configurations in a cell can easily be generated. A non-directed graph representing the adjacency relations between cells is then constructed and searched. This is called the *connectivity graph* or *adjacency graph*. The outcome of the search is a sequence of cells called a *channel*. There are two cell decomposition methods:

- approximate cell decomposition — Cells have predefined shapes (such as triangles or trapezoids) whose union is strictly included in the free space. The boundary of a cell does not characterize a discontinuity of some sort and has no physical meaning. The quad-treemethod recursively decomposes a rectangular cell into four smaller cells until the cells lie entirely in the free space or reach some resolution. These methods are sound but notincomplete.

- exact cell decomposition — The free space is decomposed into cells whose union is exactly the free space; this is a complete method. Cells in this form of decomposition are called cylinders.

Potential Fields

A local approach to motion planning. The goal configuration generates an attractive potential that pulls the robot toward it; C-obstacles generate repulsive potential that pushes the robot away. The negated gradient of the total potential is treated as an artificial force applied to the robot, considered the most promising direction.

Potential fields are very efficient but suffer from local minima. Two approaches overcome this:

- design potential functions with no local minima other than the goal configuration
- complement the basic potential field approach with mechanisms to escape from local minima

Landmark-Based Navigation

Landmark-based navigation assumes that the environment contains easily recognizable, unique*landmarks*. A landmark is modeled as a point surrounded by a *field of influence*. Within this field of influence, the robot knows its position exactly. Outside of all fields of influence, the robot has no direct position information.

Landmarks can be used to deal with uncertainty in motion so, for example, if a robot knows it is apt to move with slight uncertainty as it travels, it can aim for a central point in a field of influence, knowing that the actual cone it maps out will intersect with that field no matter how off its projections it is.

Configuration Space

For a robot with k degrees of freedom, the state or configuration of the robot can be described by k real values. These values can be considered as a point p in a k-dimensional *configuration space* of the robot.

Configuration space can be used to determine if there is a path by which a robot can move from one place to another. Real obstacles in the world are mapped to *configuration space obstacles*, and the remaining *free space* is all of configuration space except the part occupied by those obstacles.

Having made this mapping, suppose there are two points in configuration space. The robot can move between the two corresponding real world points exactly when there is a continuous path between them that lies entirely in configuration free space.

Generalized Configuration Space

The term *generalized configuration space* is used to describe systems in which other objects are included as part of the configuration. These may be movable, and their shapes may vary.

There are several ways of dealing with robot planning when there are several moving or movable objects. These are:

1. Partition the generalized configuration space into finitely many states. The planning problem then becomes a logical one, like the blocks world. No general method for partitioning space has yet been found.

2. Plan object motions first, and then motions for the robot.

3. Restrict object motions to simplify planning.

EXTENSIONS OF THE BASIC PROBLEM

Multiple Moving Objects

Obstacles may be moving. There may be multiple robots. The robots themselves may be articulated (*i.e.*, made of rigid objects connected by joints).

The first extension, and possibly the second, requires time to be explicitly considered, which gives rise to *configuration-time* space in which a path must never go back along the time axis, and some constraints on the slope and curvature of the path are given due to constraints on velocity and acceleration.

The second and third extensions yield configuration spaces of arbitrarily large dimensions. Planning can be done in a *composite configuration space* which is the cross-product of the individual configuration spaces (this is called *centralized planning*), or another method called *decoupled planning* can be used to plan the motions more or less independently and interactions are only considered in the second phase of planning.

The decoupling method fails when interaction is critical — in the example of two robots trying to switch places in a narrow corridor, for example. When dealing with articulated robots, the dimension of the configuration space grows as the number of joints.

Kinematic Constraints

A robot's motions may be restricted by kinematic constraints. There are two kinds:

Holonomic Constraints

A holonomic equality constraint is an equality relation among the parameters of the minimally-represented configuration space that can be solved for one of the parameters. Such a relation reduces the dimension of the actual configuration space of the robot by one. A set of k holonomic constraints reduces it by k. For example, a robot limited to rotating around a fixed axis has a configuration space of dimension 4 instead of 6 (since revolute joints impose 2 holonomic constraints). nonholonomic constraints

A nonholonomic equality constraint is a non-integrable equation involving the configuration parameters and their derivatives (velocity parameters). Such a constraint does not reduce the dimension of the configuration space, but instead reduces the dimension of the space of possible differential motions.

For example, a car-like robot has 3 dimensions: two for translation and one for rotation. However, the velocity of **R** is required to point along the main axis of **A**.

(This is written as - $sin\ T\ dx + cos\ T\ dy = 0$.)

The instantaneous motion of the car is determined by two parameters: the linear velocity along the main axis, and the steering angle. However, when the steering angle is non-zero, the robot changes orientation, and its linear velocity with it, allowing the robot'sconfiguration to span a three-dimensional space. Restricting the steering angle to pi/2 restricts the set of possible differential motions without changing its dimension.

Nonholonomic constraints restrict the geometry of the feasible free paths between two configurations. They are much harder to deal with in a planner than holonomic constraints.

In general, a robot with nonholonomic constraints has fewer *controllable* degrees of freedom than it has actual degrees of freedom; these are equal in a holonomic robot.

Uncertainty

A robot may have little or no prior knowledge about its workspace. The more incomplete the knowledge, the less important the role of planning. A more typical situation is when there are errors in robot control and in the initial models, but these errors are contained within bounded regions.

Grasp Planning

Many typical robot operations require the robot to grasp an object. The rest of the operation is strongly influenced by choices made during grasping.

Grasping requires positioning the gripper on the object, which requires generating a path to this position. The grasp position must be accessible, stable, and robust enough to resist some external force. Sometimes a satisfactory position can only be reached by grasping an object, putting it down, and re-grasping it. The grasp planner must choose configurations so that the grasped objects are stable in the gripper, and it should also choose operations that reduce or at least do not increase the level of uncertainty in the configuration of the object.

The object to be grasped is the *target object*. The *gripping surfaces* are the surfaces on the robotused for grasping.

There are three principal considerations in gripping an object. They are:

- **safety** — the robot must be safe in the initial and final configurations
- **reachability** — the robot must be able to reach the initial grasping configuration and, with the object in hand, reach the final configuration
- **stability** — the grasp should be stable in the presence of forces exerted on the grasped object during transfer and parts-mating motions

Example: peg placement. By tilting a peg the robot can increase the likelihood that the initial approach conditions will have the peg part way in the hole. Other solutions are chamfers (a widening hole, producing the same effect as tilting), search along the edge, and biased search (introduce bias so that search can be done in at most one motion and not two, if the error direction is not known). Computational Complexity

These theorems are taken from Latombe's book.

- **Theorem 1 (Polyhedral bodies):** Planning a free path for a robot made of an arbitrary number of polyhedral bodies connected together at some joint vertices, among a finite set of polyhedral obstacles, between any two given configurations is a PSPACE-hard problem.

- Put another way, planning a free path in a configuration space of arbitrary dimension among fixed obstacles is PSPACE-hard.

- **Theorem 2 (Joints):** In the absence of obstacles, deciding whether a planar linkage in some initial configuration can be moved so that a certain joint reaches a given point in the plane is PSPACE-hard.

- **Theorem 3 (Rectangular robots):** Planning a path in the configuration space of a multi-bodied robot consisting of all rectangles is PSPACE-hard.

- **Theorem 4 (Planar arm):** Consider a planar arm consisting of arbitrarily many links serially connected by revolute joints such that all the links are constrained to move in the plane. Planning a free path for such an arm among a finite number of polygonal obstacles, between any two given configurations is PSPACE-hard.

- **Theorem 5 (Upper bound for fixed dimension):** A free path in a configuration space of any fixed dimension m, when the free space is a set defined by n polynomial constraints of maximal degree d, can be computed by an algorithm whose time complexity is exponential in m and polynomial in both n and d. This theorem is based on reducing the planning problem to the problem of deciding the satisfiability of sentences in the first-order theory of the reals. The important observation is that the complexity of path planning increases exponentially with the dimension of the configuration space.

- **Theorem 6 (Velocity-bounding):** Planning the motion of a rigid object translating without rotation in three dimensions among arbitrarily many moving obstacles that may both translate and rotate is a PSPACE-hard problem if the velocity modulus of the object is bounded, and an NP-hard problem otherwise.

- **Theorem 7 (Uncertainty):** Planning compliant motions for a point in the presence of uncertainty, in a three-dimensional polyhedral configuration space with an arbitrarily large number of faces, is an NEXPTIME-hard problem.

Reducing Complexity

It is possible to approximate the real problem before giving it to the motion planner; it is reasonable to trade generality for improved time performance.

One way of reducing complexity is by reducing the dimension of the configuration space. This can be done by replacing the robot by the surface or volume it sweeps out when it moves along r independent axes. This corresponds to projecting the m-dimensional configuration space along r of its dimensions. The projected configuration space has only $m - r$ dimensions. (As an example, imagine a robot with an extended arm as a disk; then you don't need to worry about rotations.)

Simplification heuristics make only local plans, by breaking the problem into subproblems. For example, many localities are stereotyped situations, such as moving through a door or turning in a corridor.

ROBOT ARCHITECTURES

Brooks' Subsumption Architecture

This approach breaks the problem down into *task-achieving behaviors* (such as wandering, avoiding obstacles, or making maps) rather than decomposing it functionally (into sensing, planning, acting).

In subsumption architectures, *levels of competence* are stacked one on top of another, ranging from the lowest level (object avoidance) to higher levels for planning and map-making. Higher levels may interfere with lower levels, but each level's architecture is built, tested and completed before the next level is added. In this way the system is robust and incrementally more powerful.

Individual levels consist of augmented finite state machines connected by message-passing wires. Higher levels may inhibit signals on these wires, or replace them with their own signals; this is how they exercise control over more basic functions.

Shakey

Shakey was the forerunner of many intelligent robot projects. The Shakey system was controlled by PLANEX, which accepted goals from the user, called a STRIPS subsystem to generate plans, and then executed them *via intermediate-level actions*. These ILA's were translated into complex routines of *low-level actions* that had some error detection and correction capabilities.

After each action was executed, PLANES would execute the shortest plan subsequence that led to a goal and whose preconditions were satisfied. In this way, actions that failed would be retried and serendipity would lead to reduced effort. If no subsequence applied, PLANEX called STRIPS to make a new plan.

Situated Automata

An approach pioneered by Stan Rosenschein that begins with an explicit representation and reasoning system, but compiles it into a finite-state machine whose inputs come from the environment and whose outputs connect to effectors.

This compilation approach distinguishes between the use of explicit knowledge representation by the designers (the "grand strategy") and the use of explicit knowledge within the agent architecture (the "grand tactic"). Rosenschein's compiler generates FSMs that can be *proved* to correspond to logical propositions about the environment, provided the compiler has the correct initial state and physics.

The FLAKEY system at SRI used situated automata to navigate, run errands, and ask questions, and had no explicit representation.

Other Architectures

TCA [Simmons and Mitchell, 1989]. This architecture combines reactive systems with traditional AI planning. TCA is a distributed system with centralized control, designed to control autonomousrobots for long periods in unstructured environments, such as the surface of Mars.

Robo-SOAR [Laird *et al..*, 1980] is an extension of SOAR to a robot architecture.

PRS [Georgeff and Lansky, 1987] is a symbolic robot planning system that interleaves planning and execution. In PRS, goals represent robot behaviors rather than world states. PRS contains procedures for turning goals into subgoals or iterations of subgoals.

THE ROBOTICS TRANSFORMATION

Robotics will eventually utterly transform military affairs. Autonomous missiles have a long history and robots are increasingly used for surveillance, auxiliary tasks and armed combat (usually controlled by human operators). From a technical point of view, existing and emerging robot technologies have the potential for rapidly replacing manned platforms with superior autonomous systems in all arenas. Yet, we are still in the early days of the robotics revolution.

One of the main hurdles of the robotics transformation is the lack of integrative standards for robotics. A global standard for plug-and-play robotics would be instrumental for transforming the structure of the industry and dramatically reduce costs thus facilitating the application of robotics to all domains of modern society. Establishing a standard could be the starting point of a dramatic expansion of the robotics industry and market. The power or combination of powers that first succeeds in establishing a plug-and-play standard encompassing a sufficiently large industrial basis would surge far ahead of competitors. Leadership in robotics would quickly translate to tangible economic and military advantages. A plug-and-play standard for robotics is a truly disruptive technology.

To exemplify the application of plug-and-play robotics consider a military task force wishing to deploy robotic scouts for operations in mountainous terrain. Ideally, commanders should be able to ask a provider to quickly assemble customized units from off-the-shelf parts. The integrator would select a six-legged locomotion module from manufacturer A, a communication module from manufacturer B, a power system from manufacturer C, weapons from manufacturer

D, sensors from multiple manufactures and a robot operating system from any of a host of providers. As the hardware and software modules are connected to the system they would automatically be integrated in the plug-and-play fashion that we are used to expect from computer components, software and peripherals.

The main work of the provider would be to verify that the pre-tested modules are integrated correctly and perhaps to assist war fighters in configuring the on-board AI for the relevant mission profiles. The cost of the hexapod robot scout would be far lower than present expectations since multiple companies compete for providing parts that are produced in large volumes for a bourgeoning international and mostly civilian market. A plug-and-play standard for robotics would in summary make the robot market of tomorrow operate like the computer market of today in which relevant hardware and software quickly can be assembled for solving the task at hand. Military users would in particular benefit from that components can be replaced in the field and that the same modules can be used in multiple systems.

Standards as CATALYSERS

The right kind of technical standards may catalyse industrial transformations with paramount consequences for society, life style and economic growth. Major examples in recent times are personal computing and telecommunications. The critical standards can be de facto standards as common in the computer industry or de jure standards as predominantly in telecommunications. The effect of such disruptive standardization is to replace vertical integration with horizontal integration. Vertical integration means that a single vendor develops the total product that is offered to customers *e.g.* as Boeing develops aircraft for airline operators.

This makes for slow development, expensive products and consumer lock-in. Horizontal integration means that a vendor produces large amounts of similar components that *via* standardized interfaces can be combined into many different end-user products. End-user products consist of many commoditized components. A personal computer includes *e.g.* operating system, motherboard, graphics card, keyboard and hard drive.

All these components are produced by many manufacturers with little qualitative differentiation and are easily integrated to a complete system by an integrator company or even a consumer. Horizontal integration leads often to rapid development of inexpensive mass-market products and consumer empowerment. The transition of an industry from vertical integration to horizontal integration is therefore also often a shift from linear growth to exponential growth.

A prerequisite for the emergence of a horizontally integrated market is a foundation of components, prototypes and proof-of-concepts. The transition is catalysed by the emergence of de jure or de facto standards that enable sufficiently easy integration of total products that satisfy pent up demands. We study an example of this kind of transition and examine the present status of the robot industry with respect to standardization.

The Personal Computer Transition

Many components and functions of what we now recognize as a personal computer (PC) were in place in 1980. The Commodore PET had a screen, keyboard and a tape recorder for data storage integrated in a single unit. Other vendors had similar products. Yet sales and consumer acceptance were slow. A worldwide total of 724,000 personal computers were shipped by about two dozen companies 1980. The turning point for the personal computer was when IBM 1981 released an open PC architecture that quickly became the de facto standard. A few years later a host of firms shipped IBM PC clones largely built from third-party components. PC worldwide shipments reached 280 million units in 2009. Almost half of the homes in highly developed countries have a PC.

The key driver for the rapid PC market penetration was the inadvertent establishment of a de facto standard for hardware and software. Personal computing was before 1981 a jungle of mutually incompatible products where core hardware, peripherals and software largely were specific for each platform. Software applications had to be crafted specifically for each platform. Few applications were hence available. All this changed when the industry leader IBM rushed to launch a PC and cut some corners in the development process by using an open hardware architecture and an operating system that also was available to competitors.

Soon many competing products implemented the PC architecture that would become a de facto standard. A new software industry thrived on the expanding market for IBM PC compatible programs and transformed the PC to a ubiquitous and genuinely useful tool for work, communication and entertainment. Standardization also enabled a burgeoning industry of computer peripherals for gaming, communication, printing and many other purposes. In spite of the rapid increase in performance, ease-of-use and applications the cost of owning a PC actually decreased.

From a consumer point of view standardization is noticeable as plug-and-play performance meaning that a new piece of software and hardware works with little or no configuration once it is installed. Because of plug-and-play integration even consumers lacking specialized knowledge and training can build a PC in a few hours – feat that would have been impossible without a standard. Military forces use PCs for many different tasks and benefit also from the reduced cost of information processing that ultimately is due to the personal computing revolution. To appreciate the power of standards-driven horizontal integration, consider how expensive a modern PC would be if hardware, operating system and applications had been developed by a military contractor according to the vertically integrated business model.

Robotics

The robotics industry of today resembles the computer industry before the 1980s. Multiple providers of vertically integrated systems offer incompatible and expensive problem-specific solutions. Architecture and interfaces are often closely guarded secrets. Any modularization applies typically only to the present provider's portfolio. Customers suffer from systems that are hard to maintain and expand because of technical lock-ins. There are about 30 providers of industry robots but no integrating standard. The need for standardization is equally obvious for military, service and entertainment robots.

Yet there are no fundamental technical reasons for not implementing a generic plug-and-play standard for robotics. Robots are similar to computers in that they include sensors (input devices), actuators (output devices), information storage and processors. Robot sensors overlap partly with computer and game console sensors but are more diverse with *e.g.* radar, ultrasound and tactile sensors.

Robot actuators are also more diverse with complex arms for industry assembly robots, many types of locomotion for mobile robots and weapons for combat robots. Computer sensors and actuators are, however, also increasingly diverse and there is nothing in the broader span of robotic output and input devices that fundamentally prevents the application of plug-and-play technology. Robots must handle real-time operations but real-time operating systems and real-time applications are common also for computers and provide no insurmountable barrier against standardization.

The need for standards in robotics has often been pointed out by experts. So why is the robotics industry locked in a horizontal integration state that obviously hampers growth? One reason may be that the robotics industry, until recently, mainly catered to corporate customers that willingly pay premium prices for industrial robots that replace even more expensive human labour. The computer industry transition occurred when consumers and small businesses became an important market.

In robotics we have just recently seen a rise of consumer service robots (*e.g.* vacuum cleaners) and entertainment robots that might herald that the robotics business is close to a similar transition point as the computer industry in the beginning of the 1980s. Secondly it might have been an accidental circumstance that an industry leader launched open PC architecture at the right moment. Robotics lacks a dominating industrial player that can enforce a de facto standard.

Emerging Standards in Robotics

There are many types of standards that are relevant for robotics. The International Organization for Standardization (ISO) provides *e.g.* a. safety standard for robots in industrial environments and the NATO Standardization agency supplies standards for UAV airworthiness (STANAG 4671). However, focus on representative standardization initiatives that address the key issue of plug-and-play integration of robots. For the sake of brevity we overlook several important standardization and code reuse efforts including the Carnegie Mellon Robot Navigation Toolkit (CARMEN) and the open-source project Player. The seven emerging standards include major open-source projects, corporate initiatives, broad industrial collaborations and initiatives from the defence sector.

The Robot Operating System (ROS) is an open-source framework for mobile robots. ROS emerged 2007 from a Stanford research framework and the driving force is California-based robotics start-up Willow Garage supported by more than twenty partners. ROS includes a peer-to-peer communication system connecting multiple heterogeneous nodes and processes and means for discovering and using services provided by processors, sensors and actuators. An expandable library of packages for relevant functions such as motion tracking, planning, locomotion and grasping ensures a continuous growth of functions. The main thrust is on UNIX-based systems but a broader support of other platforms is in the pipeline. There seems to be a considerable momentum for ROS in the open-source and academic communities and a recent review of U.S. robotics mentions ROS as the most promising emerging standard.

Orca is an open-source framework aiming at promoting re-use of robotics software. It is agnostic of programming language, operating system and architecture. The Internet Communications Engine (ICE) is used as a communications layer, for setting up a registry for service discovery and for managing components. Full Linux and partial Windows and MacOS support are available. The followers are mostly academic but include a few industrial users. The architecture independence and the policy of maximizing cross-platform interoperability at the expense of efficiency makes Orca more of a managed library of re-usable software components and less of a plug-and-play standard.

Sony has since 1997 touted OPEN-R as a standard for entertainment robotics. OPEN-R is a centralized layered architecture where each component carries data describing its function, status and interfaces. Components register automatically with the central CPU thus allowing the robot to discover the present sensor, actuator and software configuration. Sony has since the launch vacillated between openness and protectiveness. This and the basically proprietary nature of OPEN-R have barred any wider industrial adoption.

The Microsoft Robotics Developer Studio (MRDS) facilitates writing programs that concurrently handles multiple sensors and actuators in real-time. MRDS includes a visual programming environment and simulation tools for testing and visualization. Developed programs can be uploaded to robots provided that they run on-board Microsoft operating systems such as Windows CE. Means for hard-ware integration are not provided. This and the limited choice of operating systems discourage industry-wide adoption. The programming and simulation tools are the main benefits and MRDS are therefore presently mainly used for research and by hobbyists. MRDS could, however, be a forerunner of a Microsoft-based generic robotics platform.

The Object Management Group (OMG) develops enterprise integration standards including the Unified Modelling Language (UML) and the Model-Driven Architecture (MDA). The over 800 members include, Hewlett-Packard, IBM, Lockheed Martin, Microsoft, Nortrop Grumman, THALES and Unisys. The Robotics Domain Task Force (DTF) of OMG aims at extending OMG standards to robotics including both software and hardware and to increase interoperability in robotics by collaborating with other standardization organizations. The OMG program is supported by leading Asian players such as the Japan Robot Association (JARA) and The (Japanese) National Institute of Advanced Industrial Science and Technology (AIST).

AIST has developed an open implementation of the OMG specifications. The OMG Robotic Technology Component Specification describes a component model consisting of a platform-independent part and several platform-specific models. The platform-specific models include a model where components communicate directly and a distributed model employing CORBA-based middleware. DTF develops presently a standard for dynamic deployment and configuration of robotic technology components that will bring it even closer to the plug-and-play objective.

Space Plug-and-Play Avionics (SPA) is a set of standards primarily intended for rapid integration of satellites. SPA is developed by the U.S. Air Force Research Laboratory and has been proposed for standardization to the American Institute for Aeronautics and Astronautics (AIAA). Each component in an SPA system includes a data sheet (xTEDS) containing a description of all commands that can be handled and all messages that can be produced by the component. The xTEDS is essentially a machine-readable manual for using the component. The components are connected with data transport means selected from a standardized set of specifications that includes the Universal Serial Bus (USB).

A new component (*e.g.* a radar sensor) that is connected to a SPA system is automatically recognized by the other modules. An on-board processor could read the xTEDS of the new component, send commands to the sensor and subscribe to data generated by the sensor. The ontology supporting interpretation of component data sheets is presently limited to satellites but could be generalized to robotics. The SPA specifications are freely available with the exception of the ontology (Common Data Dictionary) that is needed for writing and interpreting xTEDS.

Participation in standard development is, however, restricted to U.S. organizations and a few international partners working under bi-lateral agreements. From a technical point of view SPA has the potential to expand to a generic plug-and-play standard for robotics. Satellites include sensors, actuators, locomotion, navigation, on-board computing, energy management and communication just as advanced mobile robots. A broader ontology is needed and possibly an expanded set of communication specifications. The main hurdle is the restricted set of participants and the limited scope of the present effort.

The Joint Architecture for Unmanned Systems (JAUS) was initiated by the U.S. Department of Defense and is presently developed by the Society of Automotive Engineers (SAE). JAUS takes a top-down service-oriented approach and covers presently two different levels of integration. The highest level specifies communication protocols between system components such as unmanned vehicles, payload and command and control systems. The second level handles interoperability between major subsystems of a mobile robot where each subsystem is understood to be a processing platform *e.g.* locomotion, sensor array and planning module. JAUS does not specify anything about architecture or implementation but is essentially a message-based protocol. Many extensions including human-machine interfaces and manipulation services are drafted. JAUS has a large footprint in U.S. and NATO military robotics but is not applied extensively in other industrial or in academic contexts.

From this brief review of current standardization initiatives it is obvious that the robotics community is well aware of the need for standardization and that many relevant efforts are in progress. There is, however, no leading candidate for a global plug-and-play standard. No industrial player is in the position of establishing a de facto standard as IBM once did for the PC market. Sony is too proprietary and Microsoft has at best a partial solution.

Open source groups like ROS and Orca cater mostly to academics and fail to attract substantial groups of industrial users. OMG's Robotic Domain Taskforce and Space Plug-and-Play Avionics are both focused on component-level interoperability. The former is more abstract and address the global robotics industry including major players in Asia while the latter is technically more specific but limited to satellites and avionics. JAUS focuses on system-level integration and impacts mainly on U.S. and NATO military robotics.

Several of the emerging standards could be merged to form a more complete and powerful option. The JAUS high-level approach could be combined with the component and hardware oriented Space Plug-and-Play Avionics.

Few computer scientists would claim that the IBM PC-architecture is optimal or even was the best option at the onset of the PC-revolution. Yet it transformed the industry and changed society with consequences that still reverberates. Similarly we should note that a plug-and-play robotics standard does not need to be technically optimal or respect all particular interests and opinions. Many divergent forces resist standardization. Robot manufactures that are organized according to the vertical business model guard proprietary technologies.

Academics are prone to prioritize technical optimization and will find faults in any proposed standard. Different industrial segments including space, defence, service, industrial and entertainment robotics have different requirements and traditions and there will always be arguments for specialization and adaption to the idiosyncrasies of the segment. Yet all parties would benefit enormously from a common standard. A strong integrative force such as government initiative might be needed for overcoming the diverging forces and catalysing a global standard spanning all relevant domains.

Establishing a robotics standard would not remove the need for hand-crafting military systems. For computers there is a segment for hardware and software that fulfils special military requirements of security, reliability, robustness and electromagnetic compatibility. Similar concerns will extend to robot components and systems. Antennas and other electromagnetic emitters must be integrated with care and knowledge because of electromagnetic compatibility issues. Flying and sea-going robots must comply with aerodynamic and hydrodynamic constraints. In spite of these and similar concerns it is obvious that a robotics plug-and-play standard would be instrumental for reducing cost and development time for advanced autonomous military systems.

Military robots are presently mainly operator-controlled and used for surveillance, de-mining and for combat in low-intensity conflicts. Driven by a rapid evolution of generic robot technology military robots will be increasingly autonomous, capable, affordable and available. This could well trigger an arms race and destabilize the present world order by enabling an unpredictable development and spread of innovative military technology. Aggressive second and third tier powers could, as demonstrated by plentiful historical examples, lead the march into the new era. The robotics transition in military affairs appears to be inevitable but perilous and could increase the risk of armed conflicts. Ethical concerns related to the new technology are also significant.

FUTURE OF ROBOTICS

Artificial intelligence and robotics expert Rod Brooks forecasts major changes in the next 50 years. Much in the way that computers have revolutionized society, robots may take on an increasingly significant role in people's lives. As part of the Gerard Salton Lecture Series, Brooks delivered a talk yesterday entitled "Flesh and Machines: Robots and People" to discuss potential applications of intelligent robots.

Brooks, who directs the Computer Science and Artificial Intelligence Laboratory (CSAIL) at MIT, asserted that we have more in common with robots, and machines in general, than we think.

"Mankind has had a long history of retreat from 'special-ness,'" Brooks said.

Centuries ago, humans discovered that Earth was not, in fact, the center of the universe. Later, humans and animals were found to have common ancestors. DNA as the fundamental mechanism of life means that humans and yeast are somewhat similar.

"Over time men have become less special and more like technology," Brooks said. "We only have 25,000 genes — even potatoes have more than that!"

Brooks showed videos of several robots designed in his lab. In one scene, Brooks's colleague Cynthia "plays" with a robot she designed, Kismet.

"We see her moving that eraser, then the robot moving it. They're taking turns." At least, Brooks added, that's what the average observer would think. "But when we thought about it, she was doing all the work. She was giving the robot motion cues. That set us off on reading literature on child development."

Like Cynthia, mothers give their infants motion cues. They engage in activities with their children that the children cannot do by themselves, but can be trained to do with their caregiver's help.

"What the robot sees drives what it does," Brooks said.

Inside these robots exists a three-dimensional space; the robot's emotions are a point in that space. The robot uses its emotional state to generate how it reacts to certain objects, and can display emotion through facial expressions.

In another experiment suggestive of robots' similarity to children in their earliest stages of development, the lab called in various people to speak with the robot.

"When a mother interacts with her child, she generates messages through her voice: praise, attention, prohibition,and soothing are the four basic messages," Brooks said.

In the video, when one woman said, "Good job, Kismet! Look at my smile!" in an encouraging voice, the robot smiled proudly. When another said, "No, no, that's not appropriate" in a disparaging tone, Kismet lowered his head, his large ears drooping.

Although robots like Kismet don't actually understand the meanings of words, they are able to vocally replicate phonemes. As people teach various words to Kismet and Cog, another of CSAIL's robots, the robots can repeat them and identify them with their corresponding objects.

Brooks acknowledges that the development of intelligent robots is still in beginning stages, although significant progress has been made in areas like navigation. However, he said, "I think beyond navigation, robots have new possibilities which will be important."

As the world's demographics shift in the next half century, robots can be useful in fields such as manufacturing, agriculture and elderly assistance. Brooks imagines being able to roboticize large agriculture machines for the maintenance of individual plants. Such robots could do menial and time-consuming tasks like pruning and picking.

"Europe and the U.S. import low-cost labor now... But that labor may not be there in 50 years," Brooks said.

Second, robot arms could be used for fixed automation, which is particularly useful in manufacturing. Such robots would require the dexterity of a six-year-old, said Brooks. Third, he hoped that robots could be developed to provide in-home care to the elderly, who will soon comprise a much larger demographic in places like North America, Europe, Korea and Japan.

The future, however, holds many challenges to realizing certain robotic applications. "Will we accept robots?" Brooks asked the audience.

It may be hard, he explained, for humans to come to grips with machines that may equal or surpass their own capabilities. Few people want to admit that their emotions can exist within a machine.

"I'm not saying current robots have real emotions, but if they did, it would be hard for people to accept... and there would certainly be legislation against it!" Brooks said as the audience laughed.

"I liked the lecture very much," said Hugo Fierro grad. "I already took some courses on robots, but never thought about the philosophical aspect of it. I liked his predictions, although they're very futuristic."

"It was a lot of fun. I heard some very interesting and provocative ideas," said Prof. Graeme Bailey, computer science.

And as for the possibility of Brooks' vision becoming reality someday? "I hope so," Bailey said. "If one was to answer no to that, we have a somewhat dismal future for ourselves."

Chapter 2

ROBOTICS SENSORS AND TECHNOLOGY

ROBOTIC SENSING

Robotic sensing is a branch of robotics science intended to give robots sensing capabilities, so that robots are more human-like. Robotic sensing mainly gives robots the ability to see, touch, hear and move and uses algorithms that require environmental feedback.

Vision

Method

The visual sensing system can be based on anything from the traditional camera, sonar, and laser to the new technology radio frequency identification (RFID), which transmits radio signals to a tag on an object that emits back an identification code. All four methods aim for three procedures – sensation, estimation, and matching.

Image Processing

Image quality is important in applications that require excellent robotic vision. Algorithm based on wavelet transform for fusing images of different spectra and different foci improves image quality. Robots can gather more accurate information from the resulting improved image.

Usage

Visual sensors help robots to identify the surrounding and take appropriate action. Robots analyze the image of the immediate environment imported from the visual sensor. The result is compared to the ideal intermediate or end image, so that appropriate movement can be determined to reach the intermediate or final goal.

Touch

Signal Processing

Touch sensory signals can be generated by the robot's own movements. It is important to identify only the external tactile signals for accurate operations. Recent solution applies an adaptive filter to the robot's logic. It enables the robot to predict the resulting sensor signals of its internal motions, screening these false signals out. The new method improves contact detection and reduces false interpretation.

Usage

Touch patterns enable robots to interpret human emotions in interactive applications. Four measurable features — force, contact time, repetition, and contact area change — can effectively categorize touch patterns through the temporal decision tree classifier to account for the time delay and associate them to human emotions with up to 83% accuracy. The Consistency Index is applied at the end to evaluate the level of confidence of the system to prevent inconsistent reactions.

Robots use touch signals to map the profile of a surface in hostile environment such as a water pipe. Traditionally, a predetermined path was programmed into the robot. Currently, with the integration of touch sensors, the robots first acquire a random data point; the algorithm of the robot will then determine the ideal position of the next measurement according to a set of predefined geometric primitives. This improves the efficiency by 42%.

In recent years, using touch as a stimulus for interaction has been the subject of much study. In 2010, the robot seal PARO was built, which reacts to many stimuli from human interaction, including touch. The therapeutic benefits of such human-robot interaction is still being studied, but has shown very positive results.

Hearing

Signal Processing

Accurate audio sensor requires low internal noise contribution. Traditionally, audio sensors combine acoustical arrays and microphones to reduce internal noise level. Recent solutions combine also piezoelectric devices. These passive devices use the piezoelectric effect to transform force to voltage, so that the vibration that is causing the internal noise could be eliminated. On average, internal noise up to about 7dB can be reduced.

Robots may interpret strayed noise as speech instructions. Current voice activity detection (VAD) system uses the complex spectrum circle centroid (CSCC) method and a maximum signal-to-noise ratio (SNR) beamformer. Because humans usually look at their partners when conducting conversations, the VAD system with two microphones enable the robot to locate the instructional speech by comparing the signal strengths of the two microphones. Current system is able to cope with background noise generated by televisions and sounding devices that come from the sides.

Usage

Robots can perceive our emotion through the way we talk. Acoustic and linguistic features are generally used to characterize emotions. The combination of seven acoustic features and four linguistic features improves the recognition performance when compared to using only one set of features.

Acoustic Feature

- Duration
- Energy
- Pitch
- Spectrum
- Cepstral
- Voice quality
- Wavelets

Linguistic Feature

- Bag of words
- Part-of-speech
- Higher semantics
- Varia

Movement

Usage

Automated robots require a guidance system to determine the ideal path to perform its task. However, in the molecular scale, nano-robotslack such guidance system because individual molecules cannot store complex motions and programs. Therefore, the only way to achieve motion in such environment is to replace sensors with chemical reactions. Currently, a molecular spider that has one streptavidin molecule as an inert body and three catalytic legs is able to start, follow, turn and stop when came across different DNA origami. The DNA-based nano-robots can move over 100 nm with a speed of 3 nm/min.

In a TSI operation, which is an effective way to identify tumors and potentially cancer by measuring the distributed pressure at the sensor's contacting surface, excessive force may inflict a damage and have the chance of destroying the tissue. The application of robotic control to determine the ideal path of operation can reduce the maximum forces by 35% and gain a 50% increase in accuracy compared to human doctors.

Performance

Efficient robotic exploration saves time and resources. The efficiency is measured by optimality and competitiveness. Optimal boundary exploration is possible only when a robot has square sensing area, starts at the boundary, and uses the Manhattan metric. In complicated geometries and settings, a square sensing area is more efficient and can achieve better competitiveness regardless of the metric and of the starting point.

ROBOTICS TECHNOLOGY - SENSORS

Most robots of today are nearly deaf and blind. Sensors can provide some limited feedback to the robot so it can do its job. Compared to the senses and abilities of even the simplest living things, robots have a very long way to go.

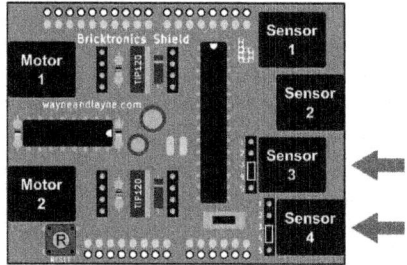

The sensor sends information, in the form of electronic signals back to the controller. Sensors also give the robot controller information about its surroundings and lets it know the exact position of the arm, or the state of the world around it.

Sight, sound, touch, taste, and smell are the kinds of information we get from our world. Robots can be designed and programmed to get specific information that is beyond what our 5 senses can tell us. For instance, a robot sensor might "see" in the dark, detect tiny amounts of invisible radiation or measure movement that is too small or fast for the human eye to see.

Here are some things sensors are used for:

Physical Property	Technology
Contact	Bump, Switch
Distance	Ultrasound, Radar, Infra Red
Light Level	Photo Cells, Cameras
Sound Level	microphones
Strain	Strain Gauges
Rotation	Encoders
Magnetism	Compasses
Smell	Chemical
Temperature	Thermal, Infra Red
Inclination	Inclinometers, Gyroscope
Pressure	Pressure Gauges
Altitude	Altimeters

Sensors can be made simple and complex, depending on how much information needs to be stored. A switch is a simple on/off sensor used for turning the robot on and off. A human retina is a complex sensor that uses more than a hundred million photosensitive elements (rods and cones). Sensors provide information to the robots brain, which can be treated in various ways. For example, we can simply *react* to the sensor output: if the switch is open, if the switch is closed, go.

Levels of Processing

To figure out if the switch is open or closed, you will need to measure the voltage going through the circuit, that's electronics. Now lets say that you have a microphone and you want to recognize a voice and separate it from noise; that's signal processing. Now you have a camera, and you want to take the pre-processed image and now you need to figure out what those objects are, perhaps by comparing them to a large library of drawings; that's computation.

Sensory data processing is a very complex thing to try and do but the robot needs this in order to have a "brain". The brain has to have analog or digital processing capabilities, wires to connect everything, support electronics to go with the computer, and batteries to provide power for the whole thing, in order to process the sensory data. Perception requires the robot to have sensors (power and electronics), computation (more power and electronics, and connectors (to connect it all).

Switch Sensors

Switches are the simplest sensors of all. They work without processing, at the electronics (circuit) level. Their general underlying principle is that of an open vs. closed circuit. If a switch is open, no current can flow; if it is closed, current can flow and be detected. This simple principle can (and is) used in a wide variety of ways.

Switch sensors can be used in a variety of ways:

- contact sensors: detect when the sensor has contacted another object (*e.g.*, triggers when a robot hits a wall or grabs an object; these can even be whiskers)
- limit sensors: detect when a mechanism has moved to the end of its range
- shaft encoder sensors: detects how many times a shaft turns by having a switch click (open/close) every time the shaft turns (*e.g.*, triggers for each turn, allowing for counting rotations)

There are many common switches: button switches, mouse switches, key board keys, phone keys, and others. Depending on how a switch is wired, it can be normally open or normally closed. This would of course depend on your robot's electronics, mechanics, and its task. The simplest yet extremely useful sensor for a robot is a "bump switch" that tells it when it's bumped into something, so it can back up and turn away. Even for such a simple idea, there are many different ways of implementation.

Light Sensors

Switches measure physical contact and light sensors measure the amount of light impacting a photocell, which is basically a resistive sensor. The resistance of a photocell is low when it is brightly illuminated, *i.e.*, when it is very light; it is high when it is dark.

In that sense, a light sensor is really a "dark" sensor. In setting up a photocell sensor, because you will need to deal with the relationship of the photocell resistance photo, and the resistance and voltage in your electronics sensor circuit. Of course since you will be building the electronics and writing the program to measure and use the output of the light sensor, you can always manipulate it to make it simpler and more intuitive. What surrounds a light sensor affects its properties. The sensor can be shielded and positioned in various ways. Multiple sensors can be arranged in useful configurations and isolate them from each other with shields.

Just like switches, light sensors can be used in many different ways:

- Light sensors can measure:
    - o light intensity (how light/dark it is)
    - o differential intensity (difference between photocells)
    - o break-beam (change/drop in intensity)
- Light sensors can be shielded and focused in different ways
- Their position and directionality on a robot can make a great deal of difference and impact

Polarized light

"Normal" light emanating from a source is non-polarized, which means it travels at all orientations with respect to the horizon. However, if there is a polarizing filter in front of a light source, only the light waves of a given orientation of the filter will pass through. This is useful because now we can manipulate this remaining light with other filters; if we put it through another filter with the same characteristic plane, almost all of it will get through.

But, if we use a perpendicular filter (one with a 90-degree relative characteristic angle), we will block all of the light. Polarized light can be used to make specialized sensors out of simple photocells; if you put a filter in front of a light source and the same or a different filter in front of a photocell, you can cleverly manipulate what and how much light you detect.

Resistive Position Sensors

We said earlier that a photocell is a resistive device. We can also sense resistance in response to other physical properties, such as *bending*. The resistance of the device increases with the amount it is bent. These bend sensors were originally developed for video game control (for example, Nintendo Powerglove), and are generally quite useful. Notice that repeated bending will wear out the sensor. Not surprisingly, a bend sensor is much less robust than light sensors, although they use the same underlying resistive principle.

Potentiometers

These devices are very common for manual tuning; you have probably seen them in some controls (such as volume and tone on stereos). Typically called *pots*, they allow the user to manually adjust the resistance. The general idea is that the device consists of a movable tap along two fixed ends. As the tap is moved, the resistance changes. As you can imagine, the resistance between the two ends is fixed, but the resistance between the movable part and either end varies as the part is moved. In robotics, *pots* are commonly used to sense and tune position for sliding and rotating mechanisms.

Biological Analogs :

- All of the sensors we described exist in biological systems
- Touch/contact sensors with much more precision and complexity in all species
- Bend/resistance receptors in muscles

Reflective Optosensors

We mentioned that if we use a light bulb in combination with a photocell, we can make a break-beam sensor. This idea is the underlying principle in reflective optosensors: the sensor consists of an emitter and a detector. Depending of the arrangement of those two relative to each other, we can get two types of sensors:

- reflectance sensors (the emitter and the detector are next to each other, separated by a barrier; objects are detected when the light is reflected off them and back into the detector)
- break-beam sensors (the emitter and the detector face each other; objects are detected if they interrupt the beam of light between the emitter and the detector)

The emitter is usually made out of a light-emitting diode (an LED), and the detector is usually a photodiode/phototransistor.

Note that these are not the same technology as resistive photocells. Resistive photocells are nice and simple, but their resistive properties make them slow; photodiodes and photo-transistors are much faster and therefore the preferred type of technology.

What can you do with this simple idea of light reflectivity? Quite a lot of useful things:

- object presence detection
- object distance detection
- surface feature detection (finding/following markers/tape)
- wall/boundary tracking
- rotational shaft encoding (using encoder wheels with ridges or black & white color)
- bar code decoding

Note, however, that light reflectivity depends on the color (and other properties) of a surface. A light surface will reflect light better than a dark one, and a black surface may not reflect it at all, thus appearing invisible to a light sensor. Therefore, it may be harder (less reliable) to detect darker objects this way than lighter ones. In the case of object distance, lighter objects that are farther away will seem closer than darker objects that are not as far away. This gives you an idea of how the physical world is partially-observable. Even though we have useful sensors, we do not have complete and completely accurate information.

Another source of noise in light sensors is ambient light. The best thing to do is subtract the ambient light level out of the sensor reading, in order to detect the actual change in the reflected light, not the ambient light. How is that done? By taking two (or more, for higher accuracy) readings of the detector, one with the emitter on, and one with it off, and subtracting the two values from each other. The result is the ambient light level, which can then be subtracted from future readings. This process is called sensor *calibration*. Of course, remember that ambient light levels can change, so the sensors may need to be calibrated repeatedly.

Break-beam Sensors

We already talked about the idea of break-beam sensors. In general, any pair of compatible emitter-detector devices can be used to produce such a sensors:

- an incandescent flashlight bulb and a photocell
- red LEDs and visible-light-sensitive photo-transistors
- or infra-red IR emitters and detectors

Shaft Encoding

Shaft encoders measure the angular rotation of an axle providing position and/or velocity info. For example, a speedometer measures how fast the wheels of a vehicle are turning, while an odometer measures the number of rotations of the wheels.

In order to detect a complete or partial rotation, we have to somehow mark the turning element. This is usually done by attaching a round disk to the shaft, and cutting notches into it. A light emitter and detector are placed on each side of the disk, so that as the notch passes between them, the light passes, and is detected; where there is no notch in the disk, no light passes.

If there is only one notch in the disk, then a rotation is detected as it happens. This is not a very good idea, since it allows only a low level of resolution for measuring speed: the smallest unit that can be measured is a full rotation. Besides, some rotations might be missed due to noise.

Usually, many notches are cut into the disk, and the light hits impacting the detector are counted.

An alternative to cutting notches in the disk is to paint the disk with black (absorbing, non-reflecting) and white (highly reflecting) wedges, and measure the reflectance. In this case, the emitter and the detector are on the same side of the disk.

In either case, the output of the sensor is going to be a wave function of the light intensity. This can then be processes to produce the speed, by counting the peaks of the waves.

Note that shaft encoding measures both position and rotational velocity, by subtracting the difference in the position readings after each time interval. Velocity, on the other hand, tells us how fast a robot is moving, or if it is moving at all. *There are multiple ways to use this measure:*

- measure the speed of a driven (active) wheel
- use a passive wheel that is dragged by the robot (measure forward progress)

We can combine the position and velocity information to do more sophisticated things:

- move in a straight line
- rotate by an exact amount

Note, however, that doing such things is quite difficult, because wheels tend to slip (effector noise and error) and slide and there is usually some slop and backlash in the gearing mechanism. Shaft encoders can provide feedback to correct the errors, but having some error is unavoidable.

Quadrature Shaft Encoding

So far, we've talked about detecting position and velocity, but did not talk about direction of rotation. Suppose the wheel suddenly changes the direction of rotation; it would be useful for the robot to detect that.

An example of a common system that needs to measure position, velocity, and direction is a computer mouse. Without a measure of direction, a mouse is pretty useless. How is direction of rotation measured?

Quadrature shaft encoding is an elaboration of the basic break-beam idea; instead of using only one sensor, two are needed. The encoders are aligned so that their two data streams coming from the detector and one quarter cycle (90-degrees) out of phase, thus the name "quadrature". By comparing the output of the two encoders at each time step with the output of the previous time step, we can tell if there is a direction change. When the two are sampled at each time step, only one of them will change its state (*i.e.*, go from on to off) at a time, because they are out of phase. Which one does it determines which direction the shaft is rotating. Whenever a shaft is moving in one direction, a counter is incremented, and when it turns in the opposite direction, the counter is decremented, thus keeping track of the overall position.

Other uses of quadrature shaft encoding are in robot arms with complex joints (such as rotary/ball joints; think of your knee or shoulder), Cartesian robots (and large printers) where an arm/rack moves back and forth along an axis/gear.

Modulation and Demodulation of Light

We mentioned that ambient light is a problem because it interferes with the emitted light from a light sensor. One way to get around this problem is to emit *modulated light*, *i.e.*, to rapidly turn the emitter on and off. Such a signal is much easier and more reliably detected by a *demodulator*, which is tuned to the particular frequency of the modulated light. Not surprisingly, a detector needs to sense several on-flashes in a row in order to detect a signal, *i.e.*, to detect its frequency. This is a small point, but it is important in writing demodulator code.

The idea of modulated IR light is commonly used; for example in household remote controls. Modulated light sensors are generally more reliable than basic light sensors. They can be used for the same purposes: detecting the presence of an object measuring the distance to a nearby object

Infra Red (IR) Sensors

Infra red sensors are a type of light sensors, which function in the infra red part of the frequency spectrum. IR sensors consist are active sensors: they consist of an emitter and a receiver. IR sensors are used in the same ways that visible light sensors are that we have discussed so far: as break-beams and as reflectance sensors.

IR is preferable to visible light in robotics (and other) applications because it suffers a bit less from ambient interference, because it can be easily modulated, and simply because it is not visible.

IR Communication

Modulated infra red can be used as a serial line for transmitting messages. This is is fact how IR modems work. Two basic methods exist:

- bit frames (sampled in the middle of each bit; assumes all bits take the same amount of time to transmit)
- bit intervals (more common in commercial use; sampled at the falling edge, duration of interval between sampling determines whether it's a 0 or 1)

Ultrasonic Distance Sensing

As we mentioned before, ultrasound sensing is based on the time-of-flight principle. The emitter produces a sonar "chirp" of sound, which travels away from the source, and, if it encounters barriers, reflects from them and returns to the receiver (microphone). The amount of time it takes for the sound beam to come back is tracked (by starting a timer when the "chirp" is produced, and stopping it when the reflected sound returns), and is used to compute the distance the sound traveled. This is possible (and quite easy) because we know how fast sound travels; this is a constant, which varies slightly based on ambient temperature.

At room temperature, sound travels at 1.12 feet per millisecond. Another way to put it that sound travels at 0.89 milliseconds per foot. This is a useful constant to remember.

The process of finding one's location based on sonar is called *echolocation*. The inspiration for ultrasound sensing comes from nature; bats use ultrasound instead of vision (this makes sense; they live in very dark caves where vision would be largely useless). Bat sonars are extremely sophisticated compared to artificial sonars; they involve numerous different frequencies, used for finding even the tiniest fast-flying prey, and for avoiding hundreds of other bats, and communicating for finding mates.

Specular Reflection

A major disadvantage of ultrasound sensing is its susceptibility to *specular reflection* (specular reflection means reflection from the outer surface of the object). While the sonar sensing principle is based on the sound wave reflecting from surfaces and returning to the receiver, it is important to remember that the sound wave will not necessarily bounce off the surface and "come right back." In fact, the direction of reflection depends on the incident angle of the sound beam and the surface.

The smaller the angle, the higher the probability that the sound will merely "graze" the surface and bounce off, thus not returning to the emitter, in turn generating a false long/far-away reading. This is often called specular reflection, because smooth surfaces, with specular properties, tend to aggravate this reflection problem. Coarse surfaces produce more irregular reflections, some of which are more likely to return to the emitter.

In summary, long sonar readings can be very inaccurate, as they may result from false rather than accurate reflections. This must be taken into account when programming robots, or a robot may produce very undesirable and unsafe behavior. For example, a robot approaching a wall at a steep angle may not see the wall at all, and collide with it!

Nonetheless, sonar sensors have been successfully used for very sophisticated robotics applications, including terrain and indoor mapping, and remain a very popular sensor choice in mobile robotics.

The first commercial ultrasonic sensor was produced by Polaroid, and used to automatically measure the distance to the nearest object (presumably which is being photographed). These simple Polaroid sensors still remain the most popular off-the-shelf sonars (they come with a processor board that deals with the analog electronics). Their standard properties include:

- 32-foot range
- 30-degree beam width
- sensitivity to specular reflection
- shortest distance return

Polaroid sensors can be combined into phased arrays to create more sophisticated and more accurate sensors.

One can find ultrasound used in a variety of other applications; the best known one is ranging in submarines. The sonars there have much more focused and have longer-range beams. Simpler and more mundane applications involve automated "tape-measures", height measures, burglar alarms, *etc.*

Machine Vision

So far, we have talked about relatively simple sensors. They were simple in terms of processing of the information they returned. Now we turn to machine vision, *i.e.*, to cameras as sensors.

Cameras, of course, model biological eyes. Needless to say, all biological eyes are more complex than any camera we know today, but, as you will see, the cameras and machine vision systems that process their perceptual information, are not simple at all! In fact, machine vision is such a challenging topic that it has historically been a separate branch of Artificial Intelligence.

The general principle of a camera is that of light, scattered from objects in the environment (those are called the *scene*), goes through an opening ("iris", in the simplest case a *pin hole*, in the more sophisticated case a *lens*), and impinging on what is called the *image plane*. In biological systems, the image plane is the *retina*, which is attached to numerous rods and cones (photosensitive elements) which, in turn, are attached to nerves which perform so-called "early vision", and then pass information on throughout the brain to do "higher-level" vision processing. As we mentioned before, a very large percentage of the human (and other animal) brain is dedicated to visual processing, so this is a highly complex endeavor.

In cameras, instead of having photosensitive rhodopsin and rods and cones, we use silver halides on photographic film, or silicon circuits in charge-coupled devices (CCD) cameras. In all cases, some information about the incoming light (*e.g.,* intensity, color) is detected by these photosensitive elements on the image plane.

In machine vision, the computer must make sense out of the information it gets on the image plane. If the camera is very simple, and uses a tiny pin hole, then some computation is required to compute the projection of the objects from the environment onto the image plane (note, they will be inverted). If a lens is involved (as in vertebrate eyes and real cameras), then more light can get in, but at the price of being focused; only objects a particular range of distances from the lens will be in focus. This range of distances is called the camera's *depth of field*.

The image plane is usually subdivided into equal parts, called *pixels*, typically arranged in a rectangular grid. In a typical camera there are 512 by 512 pixels on the image plane (for comparison, there are 120×10^6 rods and 6×10^6 cones in the eye, arranged hexagonally). Let's call the projection on the image plane the *image*.

The brightness of each pixel in the image is proportional to the amount of light directed toward the camera by the surface patch of the object that projects to that pixel. (This of course depends on the reflectance properties of the surface patch, the position and distribution of the light sources in the environment, and the amount of light reflected from other objects in the scene onto the surface patch.) As it turns out, brightness of a patch depends on two kinds of reflections, one being specular (off the surface, as we saw before), and the other being diffuse (light that penetrates into the object, is absorbed, and then re-emitted). To correctly model light reflection, as well as reconstruct the scene, all these properties are necessary.

Let us suppose that we are dealing with a black and white camera with a 512 x 512 pixel image plane. Now we have an image, which is a collection of those pixels, each of which is an intensity between white and black. To find an object in that image (if there is one, we of course don't know *a priori*), the typical first step ("early vision") is to do *edge detection, i.e.,* find all the edges. How do we recognize them? We define edges as curves in the image plane across which there is significant change in the brightness.

A simple approach would be to look for sharp brightness changes by differentiating the image and look for areas where the magnitude of the derivative is large. This almost works, but unfortunately it produces all sorts of spurious peaks, *i.e.*, noise. Also, we cannot inherently distinguish changes in intensities due to shadows from those due to physical objects. But let's forget that for now and think about noise. How do we deal with noise?

We do *smoothing*, *i.e.*, we apply a mathematical procedure called *convolution*, which finds and eliminates the isolated peaks. Convolution, in effect, applies a *filter* to the image. In fact, in order to find arbitrary edges in the image, we need to convolve the image with many filters with different orientations. Fortunately, the relatively complicated mathematics involved in edge detection has been well studied, and by now there are standard and preferred approaches to edge detection.

Once we have edges, the next thing to do is try to find objects among all those edges. *Segmentation* is the process of dividing up or organizing the image into parts that correspond to continuous objects. But how do we know which lines correspond to which objects, and what makes an object? There are several cues we can use to detect objects:

1. We can have stored *models* of line-drawings of objects (from many possible angles, and at many different possible scales!), and then compare those with all possible combinations of edges in the image. Notice that this is a very computationally intensive and expensive process. This general approach, which has been studied extensively, is called *model-based vision*.

2. We can take advantage of *motion*. If we look at an image at two consecutive time-steps, and we move the camera in between, each continuous solid objects (which obeys physical laws) will move as one, *i.e.*, its brightness properties will be conserved. This hives us a hint for finding objects, by subtracting two images from each other. But notice that this also depends on knowing well how we moved the camera relative to the scene (direction, distance), and that nothing was moving in the scene at the time. This general approach, which has also been studied extensively, is called *motion vision*.

3. We can use stereo (*i.e.*, *binocular stereopsis*, two eyes/cameras/points of view). Just like with motion vision above, but without having to actually move, we get two images, which we can subtract from each other, if we know what the *disparity* between them should be, *i.e.*, if we know how the two cameras are organized/positioned relative to each other.

4. We can use *texture*. Patches that have uniform texture are consistent, and have almost identical brightness, so we can assume they come from the same object. By extracting those we can get a hint about what parts may belong to the same object in the scene.

5. We can also use *shading* and *contours* in a similar fashion. And there are many other methods, involving *object shape* and projective invariants, *etc.*

Note that all of the above strategies are employed in biological vision. It's hard to recognize unexpected objects or totally novel ones (because we don't have the models at all, or not at the ready). Movement helps catch our attention. Stereo, *i.e.*, two eyes, is critical, and all carnivores use it (they have two eyes pointing in the same direction, unlike herbivores). The brain does an excellent job of quickly extracting the information we need for the scene.

Machine vision has the same task of doing real-time vision. But this is, as we have seen, a very difficult task. Often, an alternative to trying to do all of the steps above in order to do *object recognition*, it is possible to simplify the vision problem in various ways:

1. Use color; look for specifically and uniquely colored objects, and recognize them that way (such as stop signs, for example)

2. Use a small image plane; instead of a full 512 x 512 pixel array, we can reduce our view to much less, for example just a line (that's called a *linear CCD*). Of course there is much less information in the image, but if we are clever, and know what to expect, we can process what we see quickly and usefully.

3. Use other, simpler and faster, sensors, and combine those with vision. For example, IR cameras isolate people by body-temperature. Grippers allow us to touch and move objects, after which we can be sure they exist.

4. Use information about the environment; if you know you will be driving on the road which has white lines, look specifically for those lines at the right places in the image. This is how first and still fastest road and highway robotic driving is done.

Those and many other clever techniques have to be employed when we consider how important it is to "see" in real-time. Consider highway driving as an important and growing application of robotics and AI. Everything is moving so quickly, that the system must perceive and act in time to react protectively and safely, as well as intelligently.

Now that you know how complex vision is, you can see why it was not used on the first robots, and it is still not used for all applications, and definitely not on simple robots. A robot can be extremely useful without vision, but some tasks demand it. As always, it is critical to think about the proper match between the robot's sensors and the task.

ROBOTICS - TYPES OF ROBOTS

Ask a number of people to describe a robot and most of them will answer they look like a human. Interestingly a robot that looks like a human is probably the most difficult robot to make. Is is usually a waste of time and not the most sensible thing to model a robot after a human being. A robot needs to be above all functional and designed with qualities that suits its primary tasks. It depends on the task at hand whether the robot is big, small, able to move or nailed to the ground. Each and every task means different qualities, form and function, a robot needs to be designed with the task in mind.

Mobile Robots

Fig. : Mars Explorer images and other space robot images courtesy of NASA.

Mobile robots are able to move, usually they perform task such as search areas. A prime example is the Mars Explorer, specifically designed to roam the mars surface.

Mobile robots are a great help to such collapsed building for survivors Mobile robots are used for task where people cannot go. Either because it is too dangerous of because people cannot reach the area that needs to be searched.

Mobile robots can be divided in two categories:

Rolling Robots: Rolling robots have wheels to move around. These are the type of robots that can quickly and easily search move around. However they are only useful in flat areas, rocky terrains give them a hard time. Flat terrains are their territory.

Walking Robots: Robots on legs are usually brought in when the terrain is rocky and difficult to enter with wheels. Robots have a hard time shifting balance and keep them from tumbling. That's why most robots with have at least 4 of them, usually they have 6 legs or more. Even when they lift one or more legs they still keep their balance. Development of legged robots is often modeled after insects or crawfish.Stationary Robots

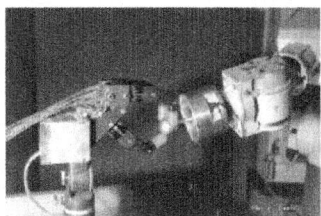

Robots are not only used to explore areas or imitate a human being. Most robots perform repeating tasks without ever moving an inch. Most robots are

'working' in industry settings. Especially dull and repeating tasks are suitable for robots. A robot never grows tired, it will perform its duty day and night without ever complaining. In case the tasks at hand are done, the robots will be reprogrammed to perform other tasks.

Autonomous Robots

Autonomous robots are self supporting or in other words self contained. In a way they rely on their own 'brains'.

Autonomous robots run a program that give them the opportunity to decide on the action to perform depending on their surroundings. At times these robots even learn new behavior. They start out with a short routine and adapt this routine to be more successful at the task they perform. The most successful routine will be repeated as such their behavior is shaped. Autonomous robots can learn to walk or avoid obstacles they find in their way. Think about a six legged robot, at first the legs move ad random, after a little while the robot adjust its program and performs a pattern which enables it to move in a direction.

Remote-control Robots

An autonomous robot is despite its autonomous not a very clever or intelligent unit. The memory and brain capacity is usually limited, an autonomous robot can be compared to an insect in that respect.

In case a robot needs to perform more complicated yet undetermined tasks an autonomous robot is not the right choice.

Complicated tasks are still best performed by human beings with real brainpower. A person can guide a robot by remote control. A person can perform difficult and usually dangerous tasks without being at the spot where the tasks are performed. To detonate a bomb it is safer to send the robot to the danger area.

Fig. : Dante 2, a NASA robot designed to explore volcanoes *via* remote control.

Virtual Robots

Virtual robots don't exits in real life. Virtual robots are just programs, building blocks of software inside a computer. A virtual robot can simulate a real robot or just perform a repeating task. A special kind of robot is a robot that searches the world wide web. The internet has countless robots crawling from site to site. These WebCrawler's collect information on websites and send this information to the search engines.

Another popular virtual robot is the chatterbot. These robots simulate conversations with users of the internet. One of the first chatterbots was ELIZA. There are many varieties of chatterbots now, including E.L.V.I.S.

BEAM Robots

BEAM is short for **B**iology, **E**lectronics, **A**esthetics and **M**echanics. BEAM robots are made by hobbyists. BEAM robots can be simple and very suitable for starters.Biology

Robots are often modeled after nature. A lot of BEAM robots look remarkably like insects. Insects are easy to build in mechanical form. Not just the mechanics are in inspiration also the limited behavior can easily be programmed in a limited amount of memory and processing power.

Electronics

Like all robots they also contain electronics. Without electronic circuits the engines cannot be controlled. Lots of Beam Robots also use solar power as their main source of energy.

Aesthetics

A BEAM Robot should look nice and attractive. BEAM robots have no printed circuits with some parts but an appealing and original appearance.

Mechanics

In contrast with expensive big robots BEAM robots are cheap, simple, built out of recycled material and running on solar energy.

ROBOTICS TECHNOLOGY - EFFECTORS

An *effector* is any device that affects the environment. Robots control their effectors, which are also known as end effectors. Effectors include legs, wheels, arms, fingers, wings and fins. Controllers cause the effectors to produce desired effects on the environment.

An *actuator* is the actual mechanism that enables the effector to execute an action. Actuators typically include electric motors, hydraulic or pneumatic cylinders, *etc*. The terms effector and actuator are often used interchangeably to mean "whatever makes the robot take an action." This is not really proper use. Actuators and effectos are not the same thing.

And we'll try to be more precise in the class. Most simple actuators control a single *degree of freedom*, *i.e.*, a single motion (*e.g.*, up-down, left-right, in-out, *etc.*). A motor shaft controls one rotational degree of freedom, for example. A sliding part on a plotter controls one translational degree of freedom. How many degrees of freedom (DOF) a robot has is going to be very important in determining how it can affect its world, and therefore how well, if at all, it can accomplish its task. Just as we said many times before that sensors must be matched to the robot's task, similarly, *effectors must be well matched to the robot's task* also.

In general, a free body in space as 6 DOF: three for translation (x,y,z), and three for orientation/rotation (roll, pitch, and yaw). We'll go back to DOF in a bit. You need to know, for a given effector (and actuator/s), how many DOF are available to the robot, as well as how many total DOF any given robot has. If there is an actuator for every DOF, then all of the DOF are controllable. Usually not all DOF are controllable, which makes robot control harder. A car has 3 DOF: position (x,y) and orientation (theta). But only 2 DOF are controllable: driving: through the gas pedal and the forward-reverse gear; steering: through the steering wheel. Since there are more DOF than are controllable, there are motions that cannot be done, like moving sideways (that's why parallel parking is hard). We need to make a distinction between what an actuator does (*e.g.*, pushing the gas pedal) and what the robot does as a result (moving forward). A car can get to any 2D position but it may have to follow a very complicated trajectory. Parallel parking requires a discontinuous trajectory w.r.t. velocity, *i.e.*, the car has to stop and go. When the number of controllable DOF is equal to the total number of DOF on a robot, it is holonomic. If the number of controllable DOF is smaller than total DOF, the robot is non-holonomic. If the number of controllable DOF is larger than the total DOF, the robot is redundant. A human arm has 7 DOF (3 in the shoulder, 1 in the elbow, 3 in the wrist), all of which can be controlled. A free object in 3D space (*e.g.*, the hand, the finger tip) can have at most 6 DOF! So there are redundant ways of putting the hand at a particular position in 3D space. This is the core of why manipulations is very hard!

Two basic ways of using effectors:

- to move the robot around =>locomotion
- to move other object around =>manipulation

These divide robotics into two mostly separate categories:

- mobile robotics
- manipulator robotics

In contrast to locomotion, where the body of the robot is moved to get to a particular position and orientation, a manipulator moves itself typically to get the *end effector* (*e.g.*, the hand, the finger, the fingertip) to the desired 3D position and orientation. So imagine having to touch a specific point in 3D space with the tip of your index finger; that's what a typical manipulator has to do.

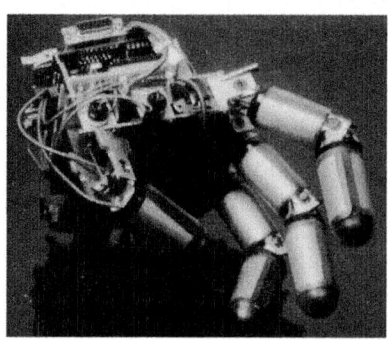

Of course, largely manipulators need to grasp and move objects, but those tasks are extensions of the basic reaching above. The challenge is to get there efficiently and safely. Because the end effector is attached to the whole arm, we have to worry about the whole arm; the arm must move so that it does not try to violate its own *joint limits* and it must not hit itself or the rest of the robot, or any other obstacles in the environment.

Thus, doing autonomous manipulation is very challenging. Manipulation was first used in tele-operation, where human operators would move artificial arms to handle hazardous materials. It turned out that it was quite difficult for human operators to learn how to tele-operate complicated arms (such as duplicates of human arms, with 7 DOF). One alternative today is to put the human arm into an exo-skeleton, in order to make the control more direct. Using joy-sticks, for example, is much harder for high DOF.

Why is this so hard? Because even as we saw with locomotion, there is typically no direct and obvious link between what the effector needs to do in physical space and what the actuator does to move it. In general, the correspondence between actuator motion and the resulting effector motion is called *kinematics*. In order to control a manipulator, we have to know its kinematics (what is attached to what, how many joints there are, how many DOF for each joint, *etc.*). We can formalize all of this mathematically, and get an equation which will tell us how to convert from, say, angles in each of the joints, to the Cartesian positions of the end effector/point. This conversion from one to the other is called computing the manipulator kinematics and *inverse kinematics*.

The process of converting the Cartesian (x,y,z) position into a set of joint angles for the arm (thetas) is called inverse kinematics. Kinematics are the rules of what is attached to what, the body structure. Inverse kinematics is computationally intense. And the problem is even harder if the manipulator (the arm) is redundant.

Manipulation involves

- trajectory planning (over time)
- inverse kinematics
- inverse dynamics
- dealing with redundancy

Manipulators are effectors. Joints connect parts of manipulators. The most common joint types are:

- rotary (rotation around a fixed axis)
- prismatic (linear movement)

These joints provide the DOF for an effector, so they are planned carefully.

Robot manipulators can have one or more of each of those joints. Now recall that any free body has 6 DOF; that means in order to get the robot's end effector to an arbitrary position and orientation, the robot requires a minimum of 6 joints. As it turns out, the human arm (not counting the hand!) has 7 DOF. That's sufficient for reaching any point with the hand, and it is also redundant, meaning that there are multiple ways in which any point can be reached.

This is good news and bad news; the fact that there are multiple solutions means that there is a larger space to search through to find the best solution. Now consider end effectors. They can be simple pointers (*i.e.*, a stick), simple 2D grippers, screwdrivers for attaching tools (like welding guns, sprayer, *etc.*), or can be as complex as the human hand, with variable numbers of fingers and joints in the fingers. Problems like reaching and grasping in manipulation constitute entire subareas of robotics and AI.

Issues include: finding grasp-points (COG, friction, *etc.*); force/strength of grasp; compliance (*e.g.*, in sliding, maintaining contact with a surface); dynamic tasks (*e.g.*, juggling, catching). Other types of manipulation, such as carefully controlling force, as in grasping fragile objects and maintaining contact with a surface (so-called *compliant motion*), are also being actively researched. Finally, dynamic manipulation tasks, such as juggling, throwing, catching, *etc.*, are already being demonstrated on robot arms.

Having talked about navigation and manipulation, think about what types of sensors (external and proprioceptive) would be useful for these general robotic tasks. *Proprioceptive* sensors sense the robot's actuators (*e.g.*, shaft encoders, joint angle sensors, *etc.*); they sense the robot's own movements. You can think of them as perceiving internal state instead of external state. External sensors are helpful but not necessary or as commonly used.

ROBOTICS TECHNOLOGY - ACTUATORS

Actuators, also known as drives, are mechanisms for getting robots to move. Most actuators are powered by pneumatics (air pressure), hydraulics (fluid pressure), or motors (electric current). Most actuation uses electromagnetic motors and gears but there have been frequent uses of other forms of actuation including NiTinOL"muscle-wires" and inexpensive Radio Control servos. To get a motor under computer control, different motor types and actuator types are used. Some of the motor types are Synchronous, Stepper, AC servo, Brushless DC servo, and Brushed DC servo.

Radio Control servos for model airplanes, cars and other vehicles are light, rugged, cheap and fairly easy to interface. Some of the units can provide very high torque speed. A Radio Control servo can be controlled from a parallel port. With one of the PC's internal timers cranked up, it is possible to control eight servos from a common parallel port with nothing but a simple interrupt service routine and a cable. In fact, power can be pulled from the disk drive power connector and the PC can run all servos directly with no additional hardware. The only down side is that the PC wastes some processing power servicing the interrupt handler.

DC Motors

The most common actuator you will use (and the most common in mobile robotics in general) is the *direct current (DC) motor*. They are simple, cheap, and easy to use. Also, they come in a great variety of sizes, to accommodate different robots and tasks. *DC motors convert electrical into mechanical energy.* They consist of permanent magnets and loops of wire inside. When current is applied, the wire loops generate a magnetic field, which reacts against the outside field of the static magnets. The interaction of the fields produces the movement of the shaft/armature.

Thus, electromagnetic energy becomes motion. As with any physical system, DC motors are not perfectly efficient, meaning that the energy is not converted perfectly, without any waste. Some energy is wasted as heat generated by friction of mechanical parts. *Inefficiencies* are minimized in well-designed (and more expensive) motors, and their performance can be brought up to the 90th percentile, but cheap motors (such as the ones you may use) can be as low as 50%. (In case you think this is very inefficient, remember that other types of effectors, such as miniature electrostatic motors, may have much lower efficiencies still.)

A motor requires a power source within its *operating voltage, i.e.,* the recommended voltage range for best efficiency of the motor. Lower voltages will usually turn the motor (but provide less power). Higher voltages are more tricky: in some cases they can increase the power output but almost always at the expense of the operating life of the motor. *E.g.,* the more you rev your car engine, the sooner it will die. When constant voltage is applied, *a DC motor draws current in the amount proportional to the work it is doing.* For example, if a robot is pushing against a wall, it is drawing more current (and draining more of its batteries) than when it is moving freely in open space.

The reason is the resistance to the motor motion introduced by the wall. If the resistance is very high (*i.e.,* the wall just won't move no matter how much the robot pushes against it), the motor draws a maximum amount of power, and stalls. This is defined as the *stall current* of the motor: the most current it can draw at its specified voltage. Within a motor's *operating current* range, the more current is used, the more *torque* or *rotational force* is produced at the shaft. In general, the strengths of the magnetic field generated in the wire loops is directly proportional to the applied current and thus the produced torque at the shaft.

Besides stall current, a motor also has its *stall torque,* the amount of rotational force produced when the motor is stalled at its operating voltage. Finally, the amount of *power* a motor generates is the product of its shaft's *rotational velocity* and its *torque.* If there is no load on the shaft, *i.e.,* the motor is spinning freely, then the rotational velocity is the highest, but the torque is 0, since no mechanism is being driven by the motor. The output power, then, is 0 also. In contrast, when the motor is stalled, it is producing maximum torque, but the rotational velocity is 0, so the output power is 0 again.

Between free spinning and stalling, the motor does useful work, and the produced power has a characteristic parabolic relationship demonstrating that the motor produces the most power in the middle of its performance range. Most DC motors have unloaded speeds in the range of 3,000 to 9,000 RPM (revolutions per minute), or 50 to 150 RPS (revolutions per second). That turns out to put them in the high-speed but low-torque category (compared to some other alternatives). For example, how often do you need to drive something very light that rotates very fast (besides a fan)? Yet that is what DC motors are naturally best at. In contrast, robots need to pull loads (*i.e.,* move their bodies and manipulators, all of which have significant mass), thus requiring more torque and less speed. As a result, the performance of a DC motor typically needs to be adjusted in that direction, through the use of *gears.*

Gearing

The force generated at the edge of a gear is equal to the product of the radius of the gear and its torque ($F = r\,t$), in the line tangential to its circumference. By combining gears with different radii, we can manipulate the amount of force/torque the mechanism generates. The relationship between the radii and the resulting torque is well defined, as follows: Suppose Gear$_1$ with radius r$_1$ turns with torque t$_1$, generating a force of t$_1$/r$_1$ perpendicular to its circumference. Now if we mesh it with Gear$_2$, with r$_2$, which generates t$_2$/r$_2$, then

$$t_1/r_1 = t_2/r_2.$$

To get the torque generated by Gear$_2$, we get:

$$t_2 = t_1\, r_2/r_1.$$

Intuitively, this means: the torque generated at the output gear is proportional to the torque on the input gear and the ratio of the two gear's radii. If r$_2$ > r$_1$, we get a bigger number, if r$_1$ > r$_2$, we get a smaller number.

If the output gear is larger than the input gear, the torque increases. If the output gear is smaller than the input gear, the torque decreases. Besides the change in torque that takes place when gears are combined, there is also a corresponding change in speed. To measure speed we are interested in the circumference of the gear,

$$C = 2\pi r.$$

Simply put, if the circumference of Gear$_1$ is twice that of Gear$_2$, then Gear$_2$ must turn twice for each full rotation of Gear$_1$. If the output gear is larger than the input gear, the speed decreases. If the output gear is smaller than the input gear, the speed increases. In summary, when a small gear drives a large one, torque is increased and speed is decreased.

Analogously, when a large gear drives a small one, torque is decreased and speed is increased. Thus, gears are used in DC motors (which we said are fast and low torque) to trade off extra speed for additional torque. Gears are combined using their teeth. The number of teeth is not arbitrary, since it is the key means of proper reduction. Gear teeth require special design so that they mesh properly. If there is any looseness between meshing gears, this is called *backlash*, the ability for a mechanism to move back \ & forth within the teeth, without turning the whole gear.

Reducing backlash requires tight meshing between the gear teeth, but that, in turn, increases *friction*. As you can imagine, proper gear design and manufacturing is complicated. To achieve "three to one gear reduction (3:1)", we apply power to a small gear (say one with 8-teeth) meshed with a large one (with 3 x 8 = 24 teeth). As a result, we have slowed down the large gear by 3 and have tripled its torque. Gears can be organized in series ("ganged"), in order to multiply their effect. For example, 2 3:1 gears in series result in a 9:1 reduction. This requires a clever arrangement of gears. Or three 3:1 gears in series can produce a 27:1 reduction. This method of multiplying reduction is the underlying mechanism that makes DC motors useful and ubiquitous.

Electronic Control of Motors

It should come as no surprise that motors require more battery power (*i.e.*, more current) than electronics (*e.g.*, 5 milliamps for the 68HC11 processor v. 100

milliamps - 1 amp for a small DC motor). Typically, specialized circuitry is required. You need to learn about H-bridges and *pulse-width modulation*there.

Servo Motors

It is sometimes necessary to be able to move a motor to a specific position. If you consider your basic DC motor, it is not built for this purpose. Motors that can turn to a specific position are called *servo motors* and are in fact constructed out of basic DC motors, by adding:

- some gear reduction
- a position sensor for the motor shaft
- an electronic circuit that controls the motor's operation

Servos are used in toys a great deal, to adjust steering on steering in RC cars and wing position in RC airplanes. Since positioning of the shaft is what servo motors are all about, most have their movement reduced to 180 degrees. The motor is driven with a waveform that specifies the desired angular position of the shaft within that range. The waveform is given as a series of pulses, within a *pulse-width modulated* signal. Thus, the width (*i.e.*, length) of the pulse specifies the control value for the motor, *i.e.*, how the shaft should turn.

Therefore, the exact width/length of the pulse is critical, and cannot be sloppy. There are no milliseconds or even microseconds to be wasted here, or the motor will behave very badly, jitter, and go beyond its mechanical limit. This limit should be checked empirically, and avoided. In contrast, the duration between the pulses is not critical at all. It should be consistent, but there can be noise on the order of milliseconds without any problems for the motor. This is intuitive: when no pulse arrives, the motor does not move, so it simply stops. As long as the pulse gives the motor sufficient time to turn to the proper position, additional time does not hurt it.

Continuous Rotation Motors

A regular DC motor can be used for continuous rotation. Furthermore, servo motors can also be retrofitted to provide continuous rotation (remember, they only to 180 otherwise), like this:

- remove mechanical limit (revert back to DC motor shaft)
- remove pot position sensor (no need to tell position)
- apply 2 resistors to fool the servo to think it is fully turning

Related Products For Drives and Actuators

Research into shape memory alloys, polymer gels and micro-mechanism devices is ongoing, and changing often. Nickel-titanium alloys were first discovered by the Naval Ordinance Laboratory decades ago and the material was termed NiTinOL. These materials have the intriguing property that they provide actuation through cycling of current through the materials.

It undergoes a 'phase change' exhibited as force and motion in the wire. At room temperature Muscle Wires are easily stretched by a small force. However, when conducting an electric current, the wire heats and changes to a much harder form that returns to the "unstretched" shape — the wire shortens in length with a usable amount of force. Nitinol can be stretched by up to eight percent of their length and will recover fully, but only for a few cycles. However when used in the three to five percent range, Muscle Wires can run for millions of cycles with very consistent and reliable performance.

ROBOTICS TECHNOLOGY - CONTROLLERS

The robot connects to a computer, which keeps the pieces of the arm working together. This computer is the controller. The controller functions as the "brain" of the robot. The controller can also network to other systems, so that the robot may work together with other machines, processes, or robots.

Given that the robot arm movement is appropriate to its application, that the arm strength and rigidity meet the payload needs and that servo drives provide the necessary speed of response and resolution, a robot controller is required to manage the arm articulations, its End Effector, and the interface with the workplace. The simplest type of control, still widely used, is "record-playback," or "lead-through". An operator positions arm articulations to desired configurations. At each desired location the articulation encoder positions are recorded in memory. Step by step, an entire work-cycle sequence is recorded. Then in playback mode the sequence is observed and modified.

As applications become more challenging, some jobs require continuous path control of an End Effector. For this action all articulations must be programmed in speeds appropriate to the particular task. This requires programming for the control of the robot. Robots today have controllers run by programs — sets of instructions written in code. The program sets limits on what the robot can do. These requirements call into play sophisticated computer-based controllers and so-called robot languages. These languages permit a kind of robot control known as hierarchical control, in which decision making by the robot takes place on several levels.

These levels are interconnected by feedback mechanisms that inform the next higher level of the status of previous actions. The advantage of a general-purpose robot arm is that it can be programmed to do many jobs. The disadvantage is that the programming tends to be a job for highly paid engineers. Even when a factory robot can perform a task more efficiently than a person, the job of programming it and setting up its workplace can be more trouble than its worth. Commotion Systems, a new California firm, is developing easier ways to program robots using pre-designed software modules. For now though, the job of "training" robots is still one of the main reasons that they are not used more.

In the future, controllers with Artificial Intelligence could allow robots to think on their own, even program themselves. This could make robots more self-reliant and independent. Angelus Research has designed an intelligent motion controller for robots that mimics the brain's three-level structure, including instinctive, behavioral, and goal levels. The controller, which can be used in unpredictable circumstances, uses a Motorola 68HC11 microprocessor.

Feedback (Closed Loop) Control

Feedback control is a means of getting a system (in our case a robot) to achieve and maintain a desired state by continuously comparing its current and desired state. The *desired state* is also called the *goal state* of the system. Note that it can be an external or internal state: for example, a thermostat monitors and controls external state (the temperature of the house), while a robot can control its internal state (*e.g.*, battery power, by recharging at proper times) or external state (*e.g.*, distance from a wall).

If the current and desired state are the same, the control system does not need to do anything. But if they are not, how does it decide what to do? That is what the design of the controller is all about. A control system must first find the difference between the current and desired states. This difference is called the *error*, and the goal of any control system is to minimize that error.

In some systems, the only information available about the error is whether it is 0 or non-0, *i.e.*, whether the current and desired states are the same. This is very little information to work with, but it is still a basis for control and can be exploited in interesting ways. Additional information about the error would be its *magnitude*, *i.e.*, how "far" the current state is from the desired state.

Finally, the last part of the error information is its*direction*, *i.e.*, is the current state too close or too far from the desires state (in whatever space it may be). Control is easiest if we have frequent feedback providing error magnitude and direction. Notice that the behavior of a feedback system oscillates around the desired state. In the case of a thermostat, the temperature oscillates around the *set point*, the desired setting.

Similarly, the robot's movement will oscillate around the desired state, which is the optimal distance from the wall. How can we decrease this oscillation? We can use a smoother/larger turning angle, and we can also use a *range* instead of a *set point* distance as the goal state. Now what happens when you have sensor error in your system? What if your sensor incorrectly tells you that the robot is far from a wall, but in fact it is not? What about vice versa? How might you address these issues? Feedback control is also called *closed loop control* because it closes the loop between the input and the output, *i.e.*, it provides the system with a measure of "progress."

Open Loop Control

The alternative to closed loop control is *open loop control*. This type of control does not require the use of sensors, since state is not fed back into the system. Such systems can operate (perform repetitive, state-independent tasks) only if they are extremely well calibrated and their environment does not change in a way that affects their performance.

We have talked about feedback control so far, but there is also an important notion of *feed forward control*. In such a system, the controller determines set points and sub-goals for itself ahead of time, without looking at actual state data.

Reactive Control

Reactive control is based on a tight loop connecting the robot's sensors with its effectors. Purely reactive systems do not use any internal representations of the environment, and do not look ahead: they *react* to the current sensory information. Thus, reactive systems use a direct mapping between sensors and effectors, and minimal, if any, state information. They consist of collections of rules that map specific *situations* to specific *actions*.

If a reactive system divides its perceptual world into *mutually exclusive* or unique situations, then only one of those situations can be triggered by any sensory input at any one time, and only one action will be activated as a result. This is the simplest form of a reactive control system. It is often too difficult to split up all possible situations this way, or it may require unnecessary encoding. Consider the case of multiple sensors: to have mutually-exclusive sensory inputs, the controller must encode rules for all possible sensory combinations.

There is an exponential number of those. This is, in fact, the robot's entire sensory space (as we defined earlier in the semester). This space then needs to be mapped to all possible actions (the action space), resulting in the complete control space for that robot. Although this mapping is done while the system is being designed, *i.e.,* not at run-time, it can be very tedious, and it results in a large look up table which takes space to encode/store in a robot, and can take time to search, unless some clever parallel look up technique is used.

In general, this complete mapping is not used in hand-designed reactive systems. Instead, specific situations trigger appropriate actions, and default actions are used to cover all other cases. Human designers can effectively reduce the sensory space to only the inputs/situations that matter, map those to the appropriate actions, and thus greatly simplify the control system. If the rules are not triggered by mutually-exclusive conditions, more than one rule can be triggered in parallel, resulting in two or more different actions being output by the system. Deciding among multiple actions or behaviors is called *arbitration*, and is in general a difficult problem. *Arbitration can be done based on:*

- fixed priority hierarchy (processes have pre-assigned priorities)
- a dynamic hierarchy (process priorities change at run-time)
- learning (process priorities may be initialized or not, and are learned at run-time, once or repeatedly/dynamically)

If a reactive system needs to support parallelism, *i.e.*, the ability to execute multiple rules at once, the underlying programming language must have the ability to *multi-task*, *i.e.*, execute several processes/pieces of code in parallel. The ability to multi-task is critical in reactive systems: if a system cannot monitor its sensors in parallel, but must go from one to another in sequence, it may miss some event, or at least the onset of an event, thus failing to react in time.

Now that we understand the building blocks of a reactive system (reactive rules coupling sensors and effectors, *i.e.*, situations and actions), we need to consider principled ways of organizing reactive controllers. We will start with the best known reactive control architecture, the Subsumption Architecture, introduced by Rod Brooks at MIT in 1985.

The Subsumption Architecture

The following are the guiding principles of the architecture:

1. Systems are built from the bottom up
2. Components are task-achieving actions/behaviors (not functional modules)
3. Components can be executed in parallel
4. Components are organized in layers, from the bottom up lowest layers handle most basic tasks
5. Newly added components and layers exploit the existing ones
6. Each component provides and does not disrupt a tight coupling between sensing and action
7. There is no need for internal models: "the world is its own best model"

Here is a rough image of how the system works: If we number the layers from 0 up, we can assume that the 0th layer is constructed, debugged, and installed first.

As layer 1 is added, layer 0 continues to function, but may be influenced by layer 1, and so on up. If layer 1 fails, layer 0 is unaffected. When layer 1 is designed, layer 0 is taken into consideration and utilized, *i.e.*, its existence is *subsumed*, thus the name of the architecture. Layer 1 can *inhibit the outputs* of layer 0 or *suppress its inputs*. Subsumption systems grow from the bottom up, and layers can keep being added, depending on the tasks of the robot. How exactly layers are split up depends on the specifics of the robot, the environment, and the task. There is no strict recipe, but some solutions are better than others, and most are derived empirically. The inspiration behind the Subsumption Architecture is the *evolutionary process*, which introduces new competencies based on the existing ones. Complete creatures are not thrown out and new ones created from scratch; instead, solid, useful substrates are used to build up to more complex capabilities.

Behavior Based Control

Behavior-based systems (BBS) use behaviors as the underlying module of the system, *i.e.*, they use a behavioral decomposition. Behaviors can vary greatly from one BBS to another, but typically have the following properties:

1. Behaviors are feedback controllers
2. Behaviors achieve specific tasks/goals (*e.g.*, avoid-others, find-friend, go-home)
3. Behaviors are typically executed in parallel/concurrently
4. Behaviors can store state and be used to construct world models/representation
5. Behaviors can directly connect sensors and effectors (*i.e.*, take inputs from sensors and send outputs to effectors)
6. Behaviors can also take inputs from other behaviors and send outputs to other behaviors (this allows for building networks)
7. Behaviors are typically higher-level than actions (go-home rather than turn-left-by-37.5-degrees)
8. Behaviors are typically closed-loop but extended in time
9. When assembled into distributed representations, behaviors can be used to look ahead but at a time-scale comparable with the rest of the behavior-based system

Behavior-based systems are not limited in the ways that reactive systems are. As a result, behavior-based systems have the following key properties:

1. The ability to react in real-time
2. The ability to use representations to generate efficient (not only reactive) behavior
3. The ability to use a uniform structure and representation throughout the system (so no intermediate layer)

The key challenge is in how representation (*i.e.*, any form of world model) can be effectively *distributed* over the behavior structure. In order to avoid the pitfalls of deliberative systems, the representation must be able to act on a time-scale that is close if not the same as the real-time parts of the system. Similarly, to avoid the pitfalls of the hybrid systems approach, the representation needs to use the same underlying behavior structure as the rest of the system. Note that behavior-based systems can have reactive components to them, *i.e.*, not every part of a behavior-based system needs to be involved with representational computation. In fact, many behavior-based systems did not use complex representations at all. As long as they use behaviors (not just rules), they are BBS.

ROBOTICS TECHNOLOGY - ARMS

The robot arm comes in all shapes and sizes and is the single most important part in robotic architecture. The arm is the part of the robot that positions the End Effector and Sensors to do their pre-programmed business. Many (but not all) resemble human arms, and have shoulders, elbows, wrists, even fingers. This gives the robot a lot of ways to position itself in its environment.

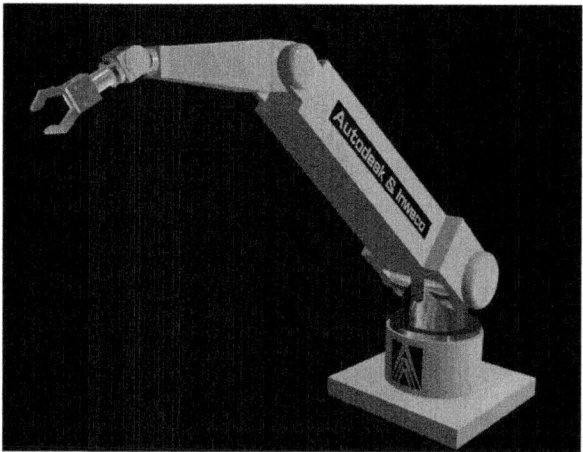

Many robots resemble human arms, and have shoulders, elbows, wrists, even fingers. This gives a robot lots of options for moving, and helps it do things in place of a human arm. In order to reach any possible point in space within its work envelope, a robot uses a total of 7 degrees of freedom. Each direction a joint can go gives an arm 1 degree. So, a simple robot arm with 3 degrees of freedom could move in 3 ways: up and down, left and right, forward and backward. Many robots of today are designed to move with these 7 degrees of freedom.

The human arm is an amazing design. It allows us to place our all-purpose end effector, the hand, where it is needed. Jointed arm robots mimic the ability of human arms to be flexible, precise, and ready for a wide variety of tasks. The jointed-arm robot has six degrees of freedom, which enable it to perform jobs that require versatility and dexterity. The design of a jointed-arm robot is similar to a human arm, but not exactly the same.

Seven Degrees of Freedom

The robotic arm is very similar to the human arm in that it has the capability of having 7 and not 6 degrees of movement, as previously thought. Out of the 7 degrees of movement, your shoulder has 3 of the arm's 7 degrees of freedom. The easiest way to explain the movement of the robotic arm is to use your own arm as an example. Hold out your arm and follow along...

First Degree: Shoulder Pitch

To find your arm's first degree of freedom: Point your entire arm straight out in front of you. Move your shoulder up and down. The up and down movement of the shoulder is called the shoulder pitch.

Second Degree: Arm Yaw

To find your arm's second degree of freedom: Point your entire arm straight out in front of you. Move your entire arm from side to side. This side to side movement is called the arm yaw.

Third Degree: Shoulder roll

To find your arm's third degree of freedom: Point your entire arm straight out in front of you. Now, roll your entire arm from the shoulder, as if you were screwing in a light bulb. This rotating movement is called a shoulder roll.

Fourth Degree: Elbow Pitch

To find your arm's fourth degree of freedom: Point your entire arm straight out in front of you. Hold your arm still, then bend only your elbow. Your elbow can move up and down. This up and down movement of the shoulder is called the shoulder pitch.

Fifth Degree: Wrist Pitch

To find your arm's fifth degree of freedom: Point your entire arm straight out in front of you. Without moving your shoulder or elbow, flex your wrist up and down. This up and down movement of the wrist is called the wrist pitch.

Sixth Degree: Wrist Yaw

To find your arm's sixth degree of freedom: Point your entire arm straight out in front of you. Without moving your shoulder or elbow, flex your wrist from side to side. The side to side movement is called the wrist yaw.

Seventh Degree: Wrist Roll

To find your arm's seventh degree of freedom: Point your entire arm straight out in front of you. Without moving your shoulder or elbow, rotate your wrist, as if you were turning a doorknob. The rotation of the wrist is called the wrist roll.

ROBOTICS TECHNOLOGY - ARTIFICIAL INTELLIGENCE

The term "artificial intelligence" is defined as systems that combine sophisticated hardware and software with elaborate databases and knowledge-based processing models to demonstrate characteristics of effective human decision making. The criteria for artificial systems include the following:

1) **functional:** the system must be capable of performing the function for which it has been designed;

2) **able to manufacture:** the system must be capable of being manufactured by existing manufacturing processes;

3) **designable:** the design of the system must be imaginable by designers working in their cultural context; and

4) **marketable:** the system must be perceived to serve some purpose well enough, when compared to competing approaches, to warrant its design and manufacture.

Robotics is one field within artificial intelligence. It involves mechanical, usually computer-controlled, devices to perform tasks that require extreme precision or tedious or hazardous work by people. Traditional Robotics uses Artificial Intelligence planning techniques to program robot behaviors and works toward robots as technical devices that have to be developed and controlled by a human engineer.

The Autonomous Robotics approach suggests that robots could develop and control themselves autonomously. These robots are able to adapt to both uncertain and incomplete information in constantly changing environments. This is possible by imitating the learning process of a single natural organism or through Evolutionary Robotics, which is to apply selective reproduction on populations of robots. It lets a simulated evolution process develop adaptive robots.

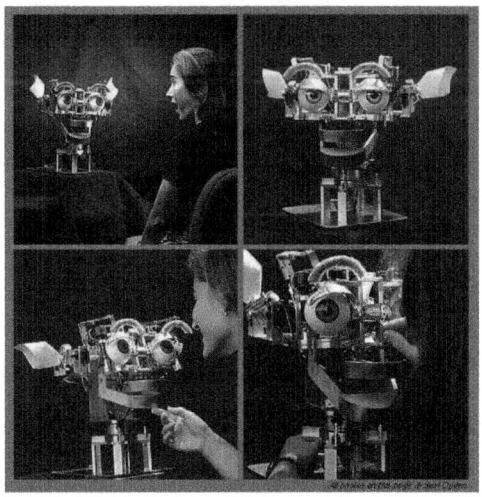

The artificial intelligence concept of the "expert system" is highly developed. This describes robot programmers ability to anticipate situations and provide the robot with a set of "if-then" rules. For example, if encountering a stairwell, stop and retreat. The more sophisticated concept is to give the robot the ability to "learn" from experience. A neural network brain equipped onto a robot will allow the robot to sample its world at random. Basically, the robot would be given some life-style goals, and, as it experimented, the actions resulting in success would be reinforced in the brain. This results in the robot devising its own rules. This is appealing to researchers and the community as it parallels human learning in lots of ways.

Artificial intelligence dramatically reduces or eliminates the risk to humans in many applications. Powerful artificial intelligence software helps to fully develop the high-precision machine capabilities of robots, often freeing them from direct human control and vastly improving their productivity. When a robot interacts with a richly populated and variable world, it uses it senses to gather data and then compare the sensate inputs with expectations that are imbedded in its world model. Therefore the effectiveness of the robot is limited by the accuracy to which its programming models the real world.

ROBOTICS TECHNOLOGY – MOBILITY

Industrial robots are rarely mobile. Work is generally brought to the robot. A few industrial robots are mounted on tracks and are mobile within their work station. Service robots are virtually the only kind of robots that travel autonomously. Research on robot mobility is extensive. The goal of the research is usually to have the robot navigate in unstructured environments while encountering unforeseen obstacles. Some projects raise the technical barriers by insisting that the locomotion involve walking, either on two appendages, like humans, or on many, like insects. Most projects, however, use wheels or tractor mechanisms. Many kinds of effectors and actuators can be used to move a robot around. Some categories are:

- legs (for walking/crawling/climbing/jumping/hopping)

- wheels (for rolling)
- arms (for swinging/crawling/climbing)
- flippers (for swimming)

Wheels

Wheels are the locomotion effector of choice. Wheeled robots (as well as almost all wheeled mechanical devices, such as cars) are built to be statically stable. It is important to remember that wheels can be constructed with as much variety and innovative flair as legs: wheels can vary in size and shape, can consist of simple tires, or complex tire patterns, or tracks, or wheels within cylinders within other wheels spinning in different directions to provide different types of locomotion properties.

So wheels need not be simple, but typically they are, because even simple wheels are quite efficient. Having wheels does not imply holonomicity. 2 or 4-wheeled robots are not usually holonomic. A popular and efficient design involves 2 differentially-steerable wheels and a passive caster. Differential steering means that the two (or more) wheels can be steered separately (individually) and thus differently. If one wheel can turn in one direction and the other in the opposite direction, the robot can spin in place. This is very helpful for following arbitrary trajectories. Tracks are often used (*e.g.*, tanks).

Legs

While most animals use legs to get around, legged locomotion is a very difficult robotic problem, especially when compared to wheeled locomotion. First, any robot needs to be stable (*i.e.*, not wobble and fall over easily). There are two kinds of stability: static and dynamic. A *statically stable* robot can stand still without falling over. This is a useful feature, but a difficult one to achieve: it requires that there be enough legs/wheels on the robot to provide sufficient static points of support.

For example, people are *not* statically stable. In order to stand up, which appears effortless to us, we are actually using active control of our balance, though nerves and muscles and tendons. This balancing is largely unconscious, but must be learned, so that's why it takes babies a while to get it right, and certain injuries can make it difficult or impossible. With more legs, static stability becomes quite simple.

In order to remain stable, *the robot's center of gravity (COG) must fall under its polygon of support*. This polygon is basically the projection between all of its support points onto the surface. So in a two-legged robot, the polygon is really a line, and the COG cannot be stably aligned with a point on that line to keep the robot upright. However, a three-legged robot, with its legs in a tripod organization, and its body above, produces a stable polygon of support, and is thus statically stable. But what happens when a statically stable robot lifts a leg and tries to move. Does its COG stay within the polygon of support? It may or may not, depending on the geometry. For certain robot geometries, it is possible (with various numbers of legs) to always stay statically stable while walking. This is very safe, but it is also very slow and energy inefficient.

A basic assumption of the static gait (statically stable gait) is that the weight of a leg is negligible compared to that of the body, so that the total center of gravity (COG) of the robot is not affected by the leg swing. Based on this assumption, the conventional static gait is designed so as to maintain the COG of the robot inside of the support polygon, which is outlined by each support leg's tip position. The alternative to static stability is *dynamic stability* which allows a robot (or animal) to be stable while moving. For example, one-legged hopping robots are dynamically stable: they can hop in place or to various destinations, and not fall over. But they cannot stop and stay standing (this is an inverse pendulum balancing problem). A statically stable robot can use dynamically-stable walking patterns, to be fast, or it can use statically stable walking. A simple way to think about this is by how many legs are up in the air during the robot's movement (*i.e.*, gait). 6 legs is the most popular number as they allow for a very stable walking gait, the tripod gait. If the same three legs move at a time, this is called the alternating tripod gait. if the legs vary, it is called the ripple gait. A rectangular 6-legged robot can lift three legs at a time to move forward, and still retain static stability.

How does it do that? It uses the so-called *alternating tripod gait*, a biologically common walking pattern for 6 or more legs. In this gait, one middle leg on one side and two non-adjacent legs on the other side of the body lift and move forward at the same time, while the other 3 legs remain on the ground and keep the robot statically stable. Roaches move this way, and can do so very quickly. Insects with more than 6 legs (*e.g.*, centipedes and millipedes), use the ripple gate. However, when they run really fast, they switch gates to actually become airborne (and thus not statically stable) for brief periods of time.

Statically stable walking is very energy inefficient. As an alternative, dynamic stability enables a robot to stay up while moving. This requires active control (*i.e.*, the inverse pendulum problem). Dynamic stability can allow for greater speed, but requires harder control. Balance and stability are very difficult problems in control and robotics, so that is why when you look at most existing robots, they will have wheels or plenty of legs (at least 6). Research robotics, of course, is studying single-legged, two legged, and other dynamically-stable robots, for various scientific and applied reasons.

Wheels are more efficient than legs. They also do appear in nature, in certain bacteria, so the common myth that biology cannot make wheels is not well founded. However, evolution favors lateral symmetry and legs are much easier to evolve, as is abundantly obvious. However, if you look at population sizes, insects are most populous animals, and they all have many more than 2 legs.

The Spider

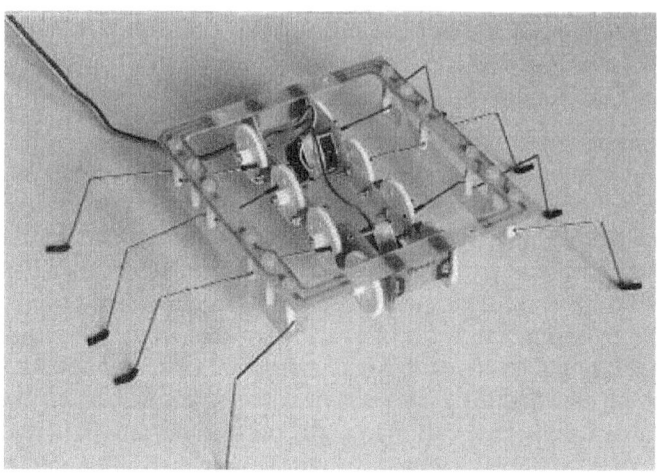

Legged Robot

In solving problems, the Spider is aided by the spring quality of its 1 mm steel wire legs. Hold one of its feet in place relative to the body and the mechanism keeps turning, the obstructed motor consuming less than 40 mA while it bends the leg. Let go and the leg springs back into shape.

As I write this, the Spider is scrambling up and over my keyboard. Some of its feet get temporarily stuck between keys, springing loose again as others push down. It has no trouble whatsoever with this obstacle, nor with any of the others on my cluttered desk - even though it is still utterly brainless.

Mobility Limits of the Spider

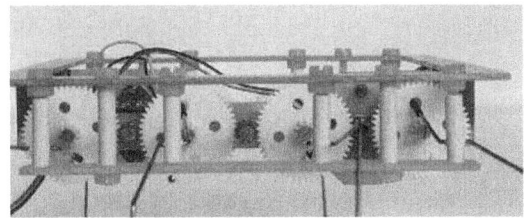

As the feet rise to a maximum of 2 cm off the floor, a cassette box is about the tallest vertical obstacle that the Spider is able to step onto. Another limitation is slope. When asked to sustain a climb angle of more than about 20 degrees, the Spider rolls over backwards. And even this fairly modest angle (extremely steep for a car, by the way) requires careful gait control, making sure that both rear legs do not lift at the same time. Improvements are certainly possible. Increasing step size would require a longer body (more distance between the legs) and thus a different gear train.

A better choice might be more legs, like 10 or 12 on a longer body, but with the same size gear wheels. That would give better traction and climbing ability. And if a third motor is allowed, one might construct a horizontal hinge in the `backbone'. Make a gear shaft the center of a nice, tight hinge joint. Then the drive train will function as before. Using the third motor and a suitable mechanism, the robot could raise its front part to step onto a tall obstacle, somewhat like a caterpillar. But turning on the spot becomes difficult.

Flying and Underwater Robots

Most robots do not fly or swim. Recently, researchers have been exploring the possibilities and problems involved with flying and swimming robots.

HETEROGENEOUS ROBOT GROUP CONTROL AND APPLICATIONS

The past decade has seen an explosion of research in robot groups. However *heterogeneous* robot groups are relatively understudied. We consider a robot group to be heterogeneous if at least one member of the group is different from the others in at least one of the following attributes:

1. mechanics,
2. sensing,
3. computing hardware or
4. nature of onboard computation.

This relatively loose condition is easily met by the system we describe in this chapter. Our experimental testbed is composed of a robot helicopter and two mobile ground robots thus making it a morphologically heterogeneous group. The advantages of using a group of robots to perform coordinated activity have been discussed extensively in the literature. Heterogeneous groups in particular, allow for the possibility of redundant solutions as well as a potentially greater degree of fault-tolerance compared to homogeneous groups.

However heterogeneous groups typically involve higher overhead in terms of system maintenance and design. At the USC Robotics Research Laboratory we have designed a heterogeneous group comprised of a robot helicopter and several mobile robots for reconnaissance and surveillance applications.

Fig. : The experimental testbed consisting of the AVATAR (Autonomous Vehicle Aerial Tracking and Reconnaissance) robot helicopter and two Pioneer AT mobile robots.

In the work described here, we focus on an experimental task motivated by an application where a single person has control and tasks the robot group at a high level. A possible application is concerned with security in urban areas where a single guard would need to control and monitor several (possibly heterogeneous) robots. In the application described here the robot group patrols an outdoor open area (a few obstacles are present on the ground) defined by a perimeter.

The perimeter is defined by a set of vertices (in GPS coordinates) which when connected by line segments form the boundary of a closed convex irregular polygon. In the work reported here the robot helicopter is flown by a human pilot for purposes of safety and speed. We have previously demonstrated control algorithms for autonomous stable hover of the helicopter, however we rely on a human pilot to transition from a stable hover at one point in space to another. Both ground robots are fully autonomous in the experiments reported here.

The robots in the experiments cover the area bounded by the perimeter and send imagery back to a human operator *via* a wireless video downlink. The operator is able to set high level goals for the ground robots. An example of a high level goal is "follow the helicopter." If a particular ground robot is given this high level goal it will stop patrolling and will follow the helicopter as it moves. The motivation behind this is to allow the ground robots to "take a closer look" at areas which the operator finds interesting based on aerial imagery.

Another example of a high level goal is "goto target." The operator can designate points of interest within the perimeter which serve as area markers for robots to periodically explore. The basic idea behind the implementation of the control programs is however to achieve robust functionality without explicit top-down planning. Rather the overall behavior of the group is the result of interacting control systems that run on the individual robots which essentially allow each robot a high degree of autonomy.

Hardware and Software Description

Our research in robotic helicopters began in 1991 with the AFV (Autonomous Flying Vehicle). We transitioned to our second robot, the AVATAR (Autonomous Vehicle Aerial Tracking And Retrieval), in 1994 and to our current robot, the second generation AVATAR (Autonomous Vehicle Aerial Tracking And Reconnaissance), in 1997. The "R" in AVATAR has changed to reflect a change in robot capabilities.

AVATAR Hardware

The current AVATAR, is based upon the Bergen Industrial Helicopter, a radio controlled (RC) model helicopter. It has a two meter diameter main rotor, is powered by a 4.0 horsepower twin cylinder gas engine and has a payload capability of approximately 10 kilograms. The helicopter has five degrees of control: main rotor lateral and longitudinal cyclic pitch, tail rotor pitch, main rotor collective pitch and engine throttle. The first three control the roll, pitch and yaw of the helicopter while the last two control its thrust. The helicopter can be controlled by a human pilot using a hand held transmitter which relays pilot control inputs to an onboard radio receiver using the 72 MHz frequency band. The receiver is connected to five actuators, one for each degree of control on the helicopter. For autonomous operation, these pilot control inputs are replaced by computer generated control inputs. A block diagram of the AVATAR system; including sensors, onboard and offboard computing resources, wireless communication links and electrical power sources. A variety of sensors are mounted on the AVATAR that provide information about the state of the helicopter as well as the environment in which it operates.

An integrated Global Positioning System/Inertial Navigation System (GPS/INS) device, consisting of a GPS receiver and an Inertial Measurement Unit (IMU), is the primary sensor used for low-level control of the helicopter. The GPS/INS provides position (latitude, longitude and altitude), velocity (horizontal and vertical), attitude (roll and pitch), heading (yaw), delta theta and delta velocity information. This particular GPS receiver can only track 4 satellites at once and consequently provides a relatively poor estimate of current latitude and longitude as compared to other available receivers.

So, a second stand alone GPS receiver is used that can track up to 12 satellites at once. This improves the standard deviations of the estimates of latitude and longitude from 4.5 meters for the 4-channel GPS unit down to 20 centimeters for the 12-channel unit. This GPS receiver is used for the high-level (guidance and navigation) control of the helicopter. A downward facing ultrasonic (sonar) transducer provides altitude information and a RPM sensor mounted on the main rotor mast measures engine speed. A downward looking color CCD camera provides visual information of the area below the AVATAR.

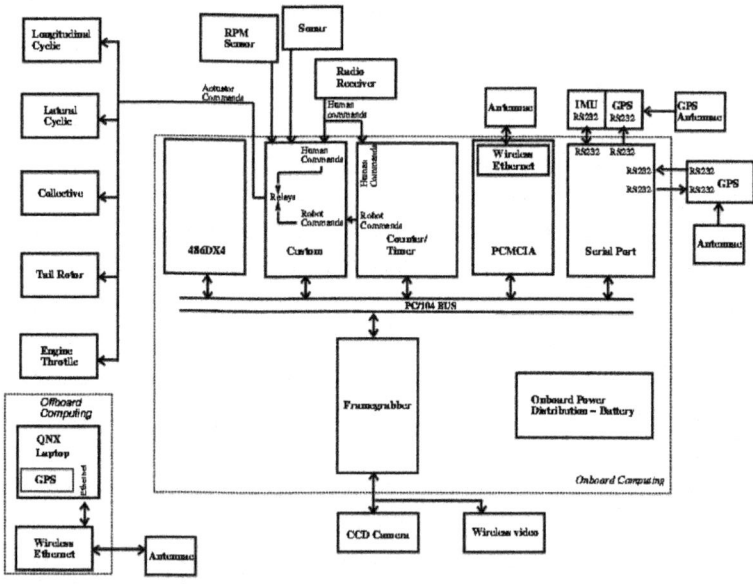

Fig. : AVATAR system block diagram.

Onboard computing needs are met using a number of commercial off the shelf (COTS) PC/104 boards and one additional custom built PC/104 board. The main processor board contains a 486DX4 CPU which runs at 133 MHz, has 16 Mbytes of RAM and 40 Mbytes of solid state disk on chip (DOC). The DOC contains both the realtime operating system (RTOS) and flight software. The 486DX4 boots up the RTOS at system powerup, executes all flight control and image processing software and provides an interface to other PC/104 boards.

These boards include a timer/counter board used both to generate actuator commands and read pilot commands, a custom built PC/104 board that allows switching between human generated and robot generated commands on an actuator by actuator basis as well as acting as an interface to the RPM sensor and sonar, a serial port board for interfacing to the GPS/INS and stand alone GPS, a color video frame grabber for the CCD camera and a PC/104 to PCMCIA interface board to allow the use of PCMCIA cards onboard the AVATAR.

A 2.4 GHz wireless Ethernet PCMCIA card provides a multiway 2.0 Mbps communication link between the AVATAR, other robots and a human using an operator control unit (OCU). The human receives grabbed video frame information and other telemetry from the AVATAR and sends high-level tasking commands to the AVATAR *via* this link. In addition, differential GPS corrections are sent from the OCU to the GPS receivers onboard the AVATAR through the wireless Ethernet to improve the GPS performance. A live video feed, provided by a one way 1.3 GHz wireless video link from the CCD camera, is displayed on a monitor.

Nickel-metal hydride (NiMH) and lithium-ion (Li-Ion) batteries supply power to the electronics. Two 10.8 volt, 4.05 amp-hour Li-Ion batteries are connected to a 50 watt PC/104 power supply board that provides +5 volts and ± 12 volts to the onboard electronics. The GPS/INS is relatively power hungry, requiring 0.8 amps at +24 volts and so a dedicated 12-volt, 3.5 amp-hour NiMH battery is connected to a DC/DC converter to produce +24 volt power. The electronics can operate for roughly an hour with this battery supply. Mission length is limited by the flight time of the helicopter on a single tank of gasoline, which is approximately 30 minutes in duration.

AVATAR Software

The operating system used is QNX; a UNIX-like, hard realtime, multitasking, extensible POSIX RTOS with a small, robust microkernel ideal for embedded systems. The microkernel handles process creation, memory management, and timer control. The flight software encompasses all remaining software running onboard the AVATAR. This currently includes (or is planned to include):

1. low-level software drivers for interfacing with sensors and actuators
2. flight control software for guidance, navigation, and control
3. learning software
4. self test software including built in test (BIT) and health checks software for increasing robot fault tolerance

5.   vision processing software running on the frame grabber

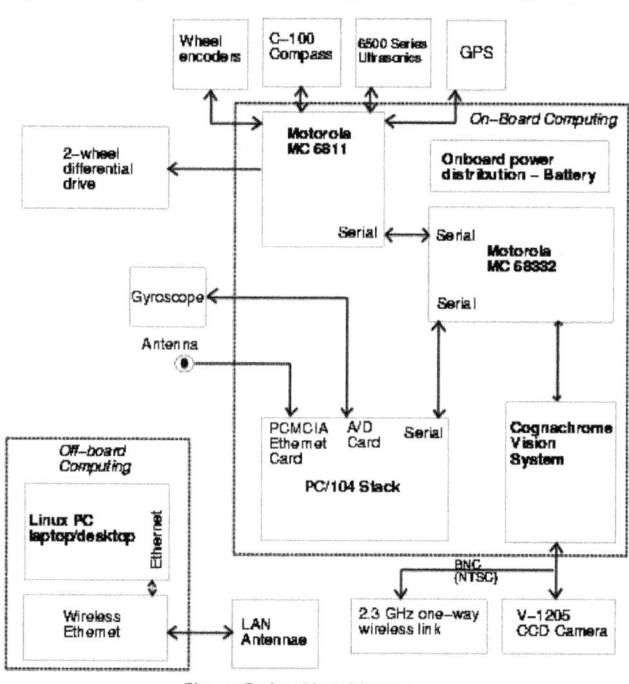

Fig. : Pioneer system block diagram.

All software is written primarily in C/C++, with assembly used only when required for speed of execution. Custom drivers have been written for the timer/counter card, the GPS/INS and GPS units, the frame grabber and the wireless Ethernet PCMCIA card. The remaining software is still under development.

Pioneer Hardware

The Pioneer AT robots used in this work are identical to each other. Each robot is four-wheeled base with skid steering. The wheels on the left are coupled mechanically as are the wheels on the right resulting in two degrees of freedom in the drive. Turning is accomplished by a speed differential between the left and right sides. Each robot has a Lithium-Ion (Li-Ion) battery pack (Two 10.8 volt, 4.05 amp-hour Li-Ion) for the electronics.

The motors are powered by a lead acid battery that allows upto 4 hours of operation on hard surfaces and approximately one hour on grass. Each robot is equipped with a ring of seven front looking sonars which are controlled by a low-level Motorola 6811 microcontroller. The wheel speeds are available through encoders. The low-level 6811 board is connected to a PC/104 stack *via* a serial connection. The Pioneer also has a vision system comprised of a camera on a pan-tilt head controlled by a Motorola 68332 board running the Congnachrome color vision system.

The main PC/104 processor board contains a 486DX4 133MHz CPU, 4 Mbytes of RAM and 40 Mbytes of solid state disk on chip which contains both the realtime operating system (RTOS) and control software. Each robot is equipped with a NovaTel GPS system connected to the PC/104 processor *via* a serial port. Additional sensors include a compass and a gyroscope. The gyroscope is connected to a A/D card on the PC/104 stack. A 2.4 GHz wireless Ethernet PCMCIA card provides a multiway 2.0 Mbps communication link between each Pioneer and the other robots. A live video feed, provided by a one-way 2.3 GHz wireless video link from the CCD camera, is displayed on a monitor. A block diagram of the Pioneer hardware.

Pioneer Software

The control system for the Pioneer AT is all written in C and runs under QNX on the PC/104 stack described above. The software includes

1. low-level software drivers for interfacing with sensors specifically software for the wireless ethernet driver and the GPS driver.
2. control software for obstacle detection and avoidance, navigation and mapping

OCU Hardware

The operator control unit is implemented using a Toshiba Tecra 510CDT laptop PC based upon a 133 MHz Pentium CPU. It has 144 Mbytes of RAM, a 4.3 Gbyte hard drive, a 12.1 inch high-res color monitor, a CD-ROM and an Ethernet connection. The QNX operating system is installed as well as the Watcom C/C++ compiler. A docking station expands the capabilities of the laptop by providing two full-length expansion slots for standard ISA and 32-bit PC Expansion Cards and one half-length 32-bit PC expansion slot. It also has a selectable bay for installing additional hardware such as CD-ROM drives or floppy drives. A 1.44M byte floppy drive has been installed in the slot.

The 510CDT has multiple functions and is used during all phases of the project; including development, integration and test. The primary purpose of the laptop is to function as a wireless OCU for communication and tasking of the robots. A 2.4 GHz wireless Ethernet device is connected to the Toshiba, providing a multiway connection between the human at the OCU and the wireless PCMCIA cards onboard the robots.

Additional functions of the 510CDT include the following: First, it provides a software development environment through the use of the QNX operating system and the Watcom C/C++ Compiler. Using this environment code is developed and tested. Also, using RCS, a QNX software version control utility, software configuration management is implemented. Second, the 4.3 Gbyte hard drive provides long-term storage capability for mission data. Third, the docking station provides an ISA slot for a GPS receiver used to produce differential GPS corrections for use by the mobile robots.

User Interface

The user interface for the system is implemented under QNX using the Phab GUI development system. The basic layout of the interface is deliberately kept simple so as to allow an inexperienced operator to quickly learn how to use it. A screenshot of the interface. The user interface allows the operator to examine telemetry from any robot, task individual robots to do specific activities (such as following, patrolling *etc.*) and monitor the location of the robots in a 2D plan view. In addition, TV monitors show the operator a live wireless video feed from each of the individual robots. Tasking is done by selecting a robot with the mouse and clicking on one of the tasks available in the form of the task list popup menu.

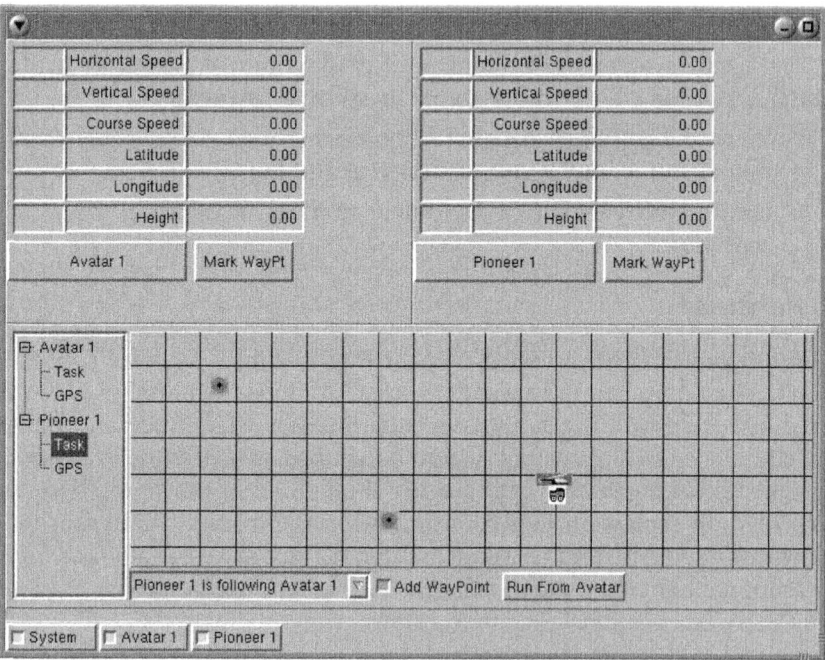

Fig. : A screenshot of the user interface.

ALGORITHMS

AVATAR Control

The AVATAR control system is implemented using a hierarchical behavior-based control system architecture. Briefly, a behavior-based control approach partitions the control problem into a set of loosely coupled computing modules (behaviors). Each behavior is responsible for a specific task and they act in parallel to achieve overall robot goals. Low-level behaviors in the hierarchy are responsible for robot functions requiring quick response while slower acting higher-level behaviors meet less time critical needs. The behavior-based control system architecture used for the AVATAR.

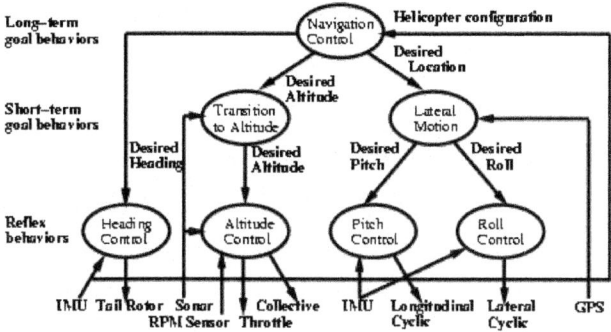

Fig. : The AVATAR Behavior-Based Control System Architecture.

At the lowest control level survival is the main priority. To this end the robot has a set of fast-acting reflex behaviors that attempt to maintain system stability by holding the craft in a hover condition. In a hover, the helicopter maintains a fixed heading and position above the ground. When the robot detects deviations the appropriate reflex returns the craft to its stable configuration. The *Heading Control* behavior attempts to hold a desired heading by using the IMU heading data to drive the tail rotor actuator. The *Altitude Control* behavior uses the sonar (ultrasonic) data and RPM sensor data to control the collective and throttle actuators. This behavior is responsible for maintaining a desired altitude above the ground. The *Pitch Control* and *Roll Control* behaviors try to hold a stable hover (zero pitch and roll orientation and rate) using the IMU angular orientation and rate data to control the longitudinal and lateral cyclic actuators. At the next level in the hierarchy, behaviors try to achieve short-term goals of the AVATAR. The *Transition to Altitude* behavior inputs a Desired Altitude to Altitude Control to move the robot to a new altitude. *Lateral Motion* generates Desired Pitch and Desired Roll commands that are given to Pitch Control and Roll Control, respectively, to move to a new lateral position.

At the top level, *Navigation Control* inputs a Desired Heading to Heading Control, a Desired Altitude to Transition to Altitude and a Desired Location to Lateral Motion to navigate the AVATAR to a new heading, altitude, latitude and longitude. Navigation Control produces these desired values based upon tasking commands received from a human using the OCU and the current Helicopter Configuration based upon sensory data. A key advantage of the behavior-based approach is the ability to create greater capabilities for the robot by layering more complex behaviors on top of previously constructed behaviors. This addition is transparent to the lower level behaviors, modulating but not destroying their underlying functionality. This allows the building and testing of control systems incrementally.

We have demonstrated completely autonomous flight on numerous occasions using this control approach with our two earlier robot helicopters, the AFV and the first generation AVATAR. A typical flight would last for five to ten minutes during which time a robot would maintain desired headings, altitudes and attitudes. We are just beginning testing with the second generation AVATAR and have demonstrated only autonomous roll control to date. The remaining actuators are currently controlled by a human pilot. As was mentioned earlier, the hardware on the AVATAR allows mixed human/robot operation.

This is accomplished through the use of the custom PC/104 hardware that switches between human and robot actuator commands on an actuator by actuator basis. We are currently learning the roll control behavior which will give the robot semi-autonomous control, able to control its roll angle but still requiring a human pilot to control the remaining four actuators. The reflex behaviors of the two earlier robots were implemented as simple proportional derivative (PD) controllers with gains laboriously tuned through a trial and error approach. This was both time intensive and dangerous.

The reflex behaviors for the current AVATAR will be implemented as hybrid fuzzy-neural controllers that are learned using an automated "teaching by showing" approach. In this approach, a controller is generated and tuned using training data gathered while a teacher operates the helicopter. No mathematical model of the dynamics of the helicopter is needed. Once the learned controllers meet desired performance criteria, control is turned over to them from the human. We have successfully applied this methodology in computer simulation, learning pitch and roll controllers capable of maintaining desired pitch and roll angles. We are in the process of validating the approach on the AVATAR. The goal is to eventually learn all of the reflex behaviors, giving the AVATAR complete autonomy.

Pioneer Control

The Pioneer robots are controlled using a behavior-based control system. The Pioneer AT robots have four wheels but only two independent controllable degrees of freedom. The wheels on either side of the robot are coupled and are always driven at the same speed.

The lowest level behavior is thus to control the speeds of the left and right wheel pairs. This is done by the commercial software available from the robot manufacturer. The *commands* to this program are generated by our control system. The Pioneer robots have a control system that is modelled as a set of interacting processes or behaviors. Sensor input is available to six behaviors. These are: regulate wheel speeds, compute heading and location, non-visual mapping, visual mapping, obstacle detection/avoidance and goal generation.

The only actuator commands generated are the wheels speeds. The laser scanner depicted in the figure is currently not used. The basic control system for each robot also accepts inputs from the basestation in the form of user commands (high level tasks) or events (such as alarms that are triggered by external entities).

The "regulate left and right wheel speeds" behavior is provided by the robot manufacturer. The computation of location and orientation. This computation is basically a Kalman filter operating on five sensor readings. These are the left and right encoders, the compass, the gyroscope and the GPS. This behavior enables a ground robot to maintain an accurate estimate of its location and orientation using an optimal combination of these sensors.

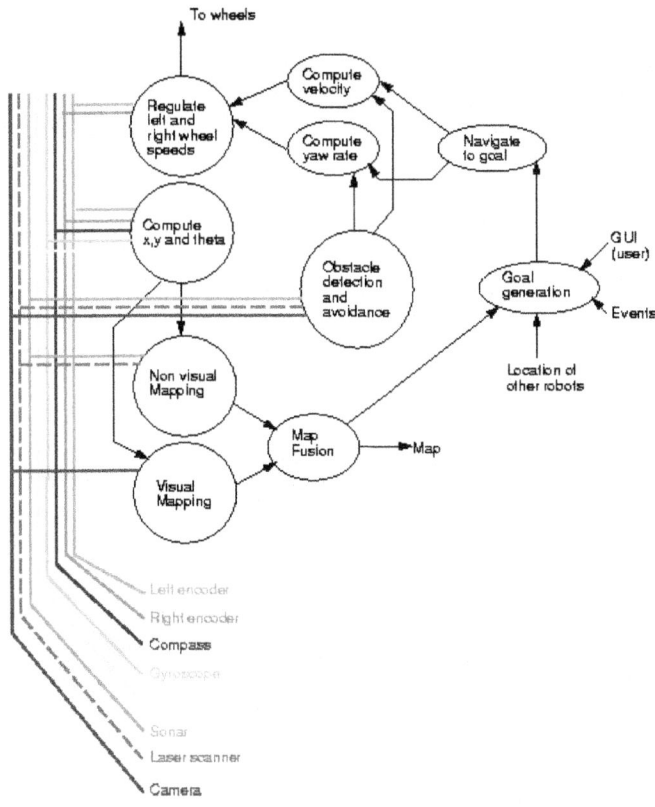

Fig. : The Pioneer behavior-based control system.

The particular form of the filter used allows different sensor readings to be collected at different frequencies. Since the filter has inputs from the odometric sensors as well as GPS, the robot is able to keep track of its location and orientation even when GPS is temporarily unavailable (as often happens in urban areas near tall buildings). The obstacle detection and avoidance behavior currently uses the front sonar ring on the robot to roughly locate the angular and radial position of obstacles in a robot-centered reference frame. This information is made available to the behaviors responsible for computing the yaw rate and the velocity of the robot.

The "navigate to goal" behavior is basically a servo-controller on the error in desired orientation and location. The error is computed as a difference between the estimated and actual values. In the absence of obstacles the robot would dead-reckon to the desired goal location. The velocity and yaw rate computation weighs the output from the "obstacle detection and avoidance" behavior more than the "navigate to goal behavior" in order that the robots stay safe. The "goal generation" behavior is currently implemented as a FIFO queue.

Inputs from the mapping behavior are timestamped and queued to a single execution list. The head of this execution list is sent as a goal to the "navigate to goal" behavior. Only one goal is sent to the "navigate to goal" behavior at any time. The decision on when to activate the next goal is made by the "goal generation" behavior which estimates the degree of progress of the prior goal. Certain events and user inputs can pre-empt the queue. Any input from these channels is treated as a LIFO queue or a stack. This allows the user to interrupt the robot in the midst of a task.

Results with Test Tasks

The system described here has been tested with two simple tasks to date. The operator initializes the system with a set of GPS coordinates that define the perimeter of interest. The robot locations are registered to a common coordinate frame which is displayed to the operator on the user interface. The first task used for testing was a cooperative patrolling activity by the robots. The helicopter was flown by the human pilot and the Pioneer ground vehicles followed it on the ground using GPS. The only input from the operator is at the start of the task. When the helicopter is fully autonomous we expect to be able to load a flight plan into it (in the form of several *via* points) and have the rest of the robots follow in loose formation.

The second task we have used for testing involves the automatic retasking of the robots by simulated "alarms" on the area perimeter. The robots navigate to the alarm scene to provide live video while maintaining basic patrolling functionality. When an alarm goes off (alarms are simulated by user keyboard input to the system), the robot nearest to the alarm disengages the patrolling behavior and navigates to the site of the alarm. Since this is a reconnaissance style scenario the robot simply returns a close-up view of the scene of the alarm to the operator in the form of a live video feed. The other robot continues patrolling.

Related Literature

For automated full-size helicopter control both model-based approaches have been used. The two approaches are merged by combining neural networks with feedback linearization. Similarly, for smaller radio controlled (RC) model helicopters and used a model-based technique while and used model-free (fuzzy logic). What separates our work from the model-based approaches is its model-free nature and online adaptive capability.

Early research in robotic control during the 1960s and 1970s was strongly influenced by the work at that time of the artificial intelligence (AI) community. AI research developed a strong dependence upon the use of abstract world models and deliberative reasoning methods. These techniques were applied to robotic control. In this approach, the control problem is attacked sequentially. That is, a robot first senses, perceives, and models its environment; then it reasons, plans and acts within the environment. Shakey, a mobile robot built in the late 1960s at the Stanford Research Institute, is representative of this AI approach for control.

In contrast to the AI approach, the behavior-based methodology solves the control problem in a parallel fashion. In a behavior-based control architecture, a controller is partitioned into a set of simpler, loosely coupled computing modules (behaviors). Each behavior is responsible for achieving a specific control task, extracts from the environment only the information required to complete its particular task, and interacts concurrently with other behaviors in order to achieve the robot's overall goals. The use of explicit abstract representational knowledge is avoided in the generation of a response by a behavior.

In an introduction to the principles and design of behavior-based autonomous robots is given as is a survey of numerous systems implemented in hardware. The pioneering work in behavior-based systems is due to Brooks, in which a purely reactive behavior-based approach is presented. Individual behaviors in the architecture are implemented as augmented finite state machines and higher layer behaviors in the architecture can subsume the actions of lower layer behaviors. Mataric extended the purely reactive subsumption architecture by integrating qualitative maps into the approach.

Mataric further extends the subsumption architecture to multiple robot group behavior with her Nerd Herd. She specifies a set of basis behaviors such as *following* where one robot follows another or *homing* where each robot attempts to return to a common home base. These basis behaviors are then combined to give more complex social interactions such as flocking or foraging. Payton *et al..* integrated fault tolerance into the behavior-based approach. Arkin has developed a hybrid deliberative and reactive approach called the Autonomous Robot Architecture (AuRA). AuRA allows the integration of behavioral, perceptual and a priori environmental knowledge into a common framework *via* a deliberative hierarchical planner, based upon traditional AI techniques, and a reactive controller, based upon schema theory.

AUTONOMOUS MOBILE ROBOT AND HARDWARE

My objectives for my recent robot design were fairly modest. I wanted to build a robot that could cruise on its own, avoid obstacles, escape from inadvertent collisions, and track a light source. I knew that if I could meet such objective other more complex behaviors would be possible (*e.g.*, self-docking on low power). There certainly many commercial robots on the market that could have met my requirements. But I decided that my best bet would be to roll my own. I wanted to keep things simple, and I wanted to fully understand the sensors and controls for behavioral autonomous operation. The TOMBOT is the fruit of that labor. A colleague came up with the name TOMBOT in honor of its inventor, and the name kind of stuck.

TOMBOT

TOMBOT Graphics Display
Content on Power Up

Fig. : a — The complete TOMBOT design. **b** — The graphics display is nice feature.

In this series of articles, I'll present lessons learned and describe the hardware/software design process.

Design Basics

The TOMBOT robot is certainly minimal, no frills: two continuous-rotation, variable-speed control servos; two IR (850 nm) analog distance measurement sensors (4- to 30-cm range); two CdS photoconductive cells with good lux response in visible spectrum; and, finally, a front bumper (switch-activated) for collision detection. The platform is simple: servos and sensors on the left and right side of two level platforms. The bottom platform houses bumper, batteries, and servos. The top platform houses sensors and microcontroller electronics. The back part of the bottom platform uses a central skid for balance between the two servos.

Given my background as a Microchip Developer and Academic Partner, I used a Microchip Technology PIC32 microcontroller, a PICkit 3 programmer/debugger, and a free Microchip IDE and 32-bit complier for TOMBOT.

It was a real thrill to design and build a minimal capability robot that can — with stacking programming behaviors — emulate some "intelligence." TOMBOT is still a work in progress, but I recently had the privilege of demoing it to a first grade class in El Segundo, CA, as part of a Science Technology Engineering and Mathematics (STEM) initiative. The results were very rewarding, but more on that later.

Behavioral Programming

A control system for a completely autonomous mobile robot must perform many complex information-processing tasks in real time, even for simple applications. The traditional method to building control systems for such robots is to separate the problem into a series of sequential functional components. An alternative approach is to use behavioral programming. The technique was introduced by Rodney Brooks out of the MIT Robotics Lab, and it has been very successful in the implementation of a lot of commercial robots, such as the popular Roomba vacuuming. It was even adopted for space applications like NASA's Mars Rover and military seekers.

Programming a robot according to behavior-based principles makes the program inherently parallel, enabling the robot to attend simultaneously to all hazards it may encounter as well as any serendipitous opportunities that may arise. Each behavior functions independently through sensor registration, perception, and action. In the end, all behavior requests are prioritized and arbitrated before action is taken. By stacking the appropriate behaviors, using arbitrated software techniques, the robot appears to show (broadly speaking) "increasing intelligence." The TOMBOT modestly achieves this objective using selective compile configurations to emulate a series of robot behaviors (*i.e.*, Cruise, Home, Escape, Avoid, and Low Power). A simple model illustration of a behavior program.

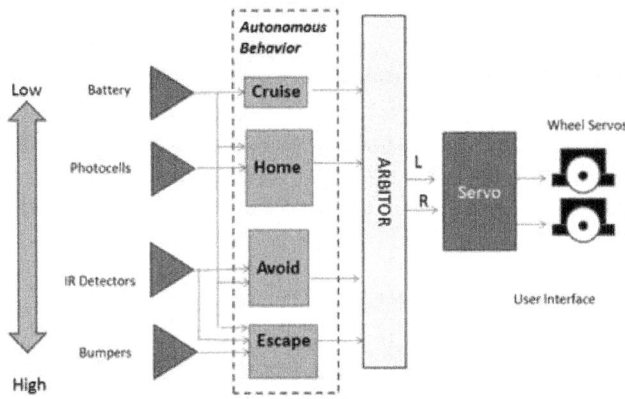

Fig. : Behavior program.

Debugging a mobile platform that is executing a series of concurrent behaviors can be daunting task. So, to make things easier, I implemented a complete remote control using a wireless link between the robot and a PC. With this link, I can enable or disable autonomous behavior, retrieve the robot sensor status and mode of operations, and curtail and avoid potential robot hazard. In addition to this, I implemented some additional operator feedback using a small graphics display, LEDs, and a simple sound buzzer. Note the TOMBOT's power-up display in **Photo 1b**. We take Robot Boot Camp*very seriously*.

Minimalist System

As you can see in the robot's block diagram, the TOMBOT is very much a minimalist system with just enough components to demonstrate autonomous behaviors: Cruise, Escape, Avoid, and Home. All these behaviors require the use of left and right servos for autonomous maneuverability.

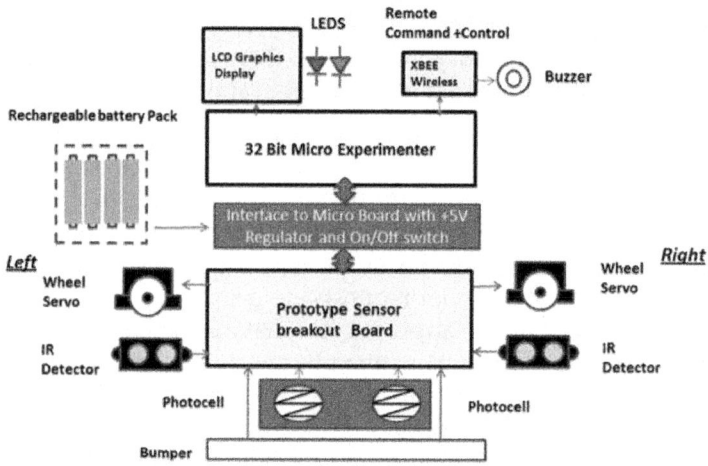

Fig. : The TOMBOT system.

The Cruise behavior just keeps the robot in motion in lieu of any stimulus. The Escape behavior uses the bumper to sense a collision and then 180° spin with reverse. The Avoid behavior makes use of continuous forward-looking IR sensors to veer left or right upon approaching a close obstacle. The Home behavior utilizes the front optical photocells to provide robot self-guidance to a strong light highly directional source. It all should add up to some very distinct "intelligent" operation. The basic sensor and electronic layout.

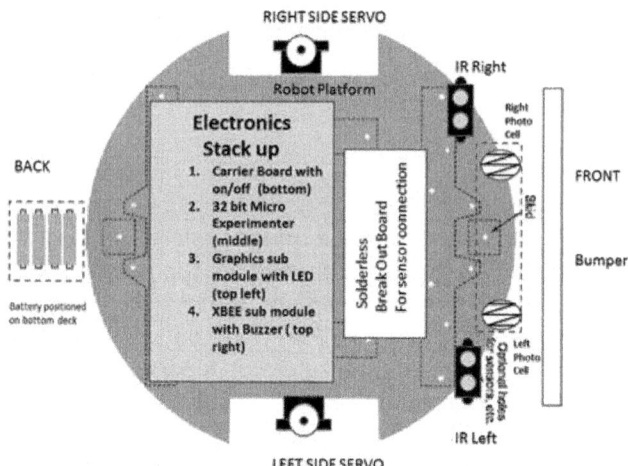

Fig. : Basic sensor and electronic layout.

TOMBOT Assembly

The TOMBOT uses the low-cost robot platform (ArBot Chassis) and wheel set (X-Wheel assembly) from Budget Robotics.

Budget Robotics ArBot Chassis

Budget Robotics X-Wheel Assembly

ArBot Assembly (top removed)
with Servos and X-Wheel Pair

Fig. : The platform and wheel set.

A picture is worth a thousand words. **Photo 2** shows two views of the TOM-BOT prototype.

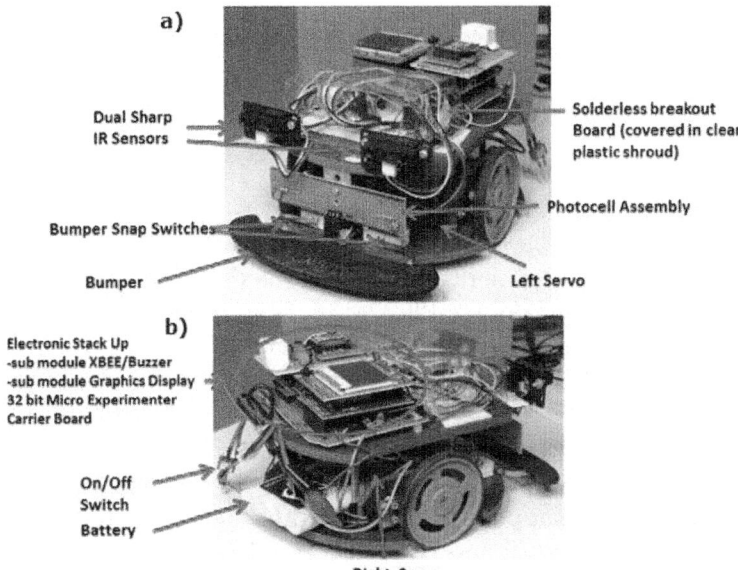

a)

Dual Sharp
IR Sensors

Solderless breakout
Board (covered in clear
plastic shroud)

Photocell Assembly

Bumper Snap Switches

Bumper   Left Servo

b)

Electronic Stack Up
-sub module XBEE/Buzzer
-sub module Graphics Display
32 bit Micro Experimenter
Carrier Board

On/Off
Switch

Battery

Right Servo

Fig. : The TOMBOT's Sharp IR sensors, photo assembly, and more.
b: The battery pack, right servo, and more.

Just below them is the photocell assembly. It is a custom board with dual CdS GL5528 photoconductive cells and 2.2-kΩ current-limiting resistors. Below this is a bumper assembly consisting of two SPDT Snap-action switches with lever (All Electronics Corp. CAT# SMS-196, left and right) fixed to a custom pre-fab plastic front bumper. Also shown is the solderless breakout board and left servo. The rechargeable battery pack that resides on the lower base platform and associated power switch.

The electronics stack is visible. Here the XBee/Buzzer and graphics card modules residing on the 32-bit Experimenter. The Experimenter is plugged into a custom carrier board that allows for an interconnection to the solderless breakout to the rest of the system. Finally, note that the right servo is highlighted. The total TOMBOT package is not ideal; but remember, I'm talking about a prototype, and this particular configuration has held up nicely in several field demos.

The same snap-action switches with extended levers are bent and fashioned to interconnect a bumper assembly as shown.

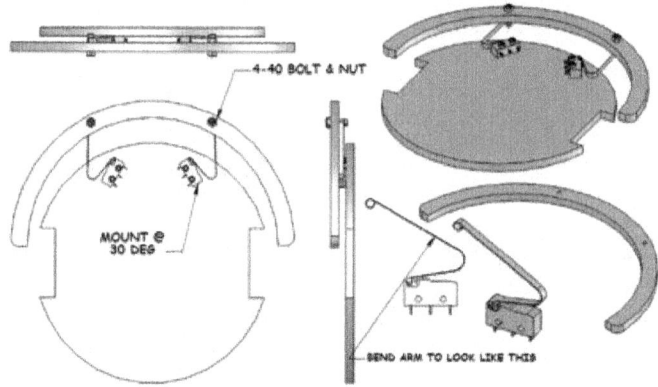

Fig. : Second-generation bumper assembly.

TOMBOT Electronics

A 32-bit Micro Experimenter is used as the CPU. This board is based the high-end Microchip Technology PIC32MX695F512H 64-pin TQFP with 128-KB RAM, 512-KB flash memory, and an 80-MHz clock. I did not want to skimp on this component during the prototype phase. In addition the 32-bit Experimenter supports a 102 × 64 monographic card with green/red backlight controls and LEDs. Since a full graphics library was already bundled with this Experimenter graphics card, it also represented good risk reduction during prototyping phase.

The Experimenter supports six basic board-level connections to outside world using JP1, JP2, JP3, JP4, BOT, and TOP headers. A custom carrier board interfaces to the Experimenter *via* these connections and provides power and signal connection to the sensors and servos. The custom carrier accepts battery voltage and regulates it to +5 VDC. This +5 V is then further regulated by the Experimenter to its native +3.3-VDC operation. The solderless breadboard supports a resistor network to sense a +9-V battery voltage for a +3.3-V PIC processor. The breadboard also contains an LM324 quad op-amp to provide a buffer between +3.3-V logic of the processor and the required +5-V operation of the servo. The schematic diagram of the electronics.

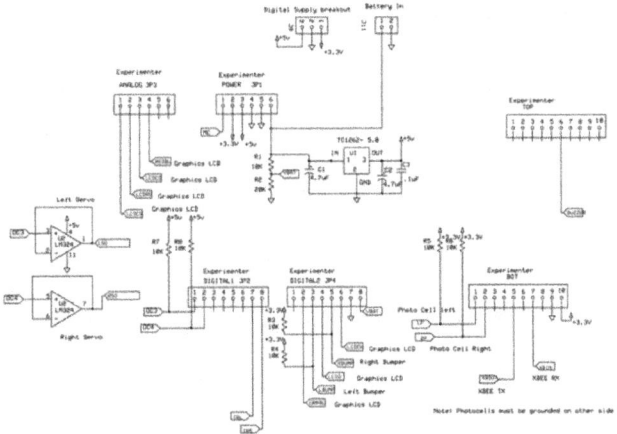

Fig. : The design's circuitry.

A custom card for the XBee radio carrier and buzzer was built that plugs into the Experimenter's TOP and BOT connections. **Photo 3** shows the modules and the carrier board. The robot uses a rechargeable 1,600-mAH battery system (typical of mid-range wireless toys) that provides hours of uninterrupted operation.

Stack Up Modules

Graphics   32 bit Experimenter   XBEE/Buzzer

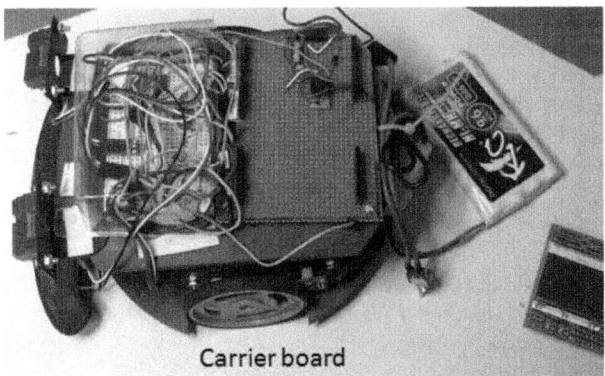

Carrier board

Fig. : The modules and the carrier board.

PIC32 On-Chip Peripherals

The major PIC32 peripheral connection for the Experimenter to rest of the system is shown. The TOMBOT uses PWM for servo, UART for XBee, SPI and digital for LCD, analog input channels for all the sensors, and digital for the buzzer and bumper detect. The key peripheral connection for the Experimenter to rest of the system.

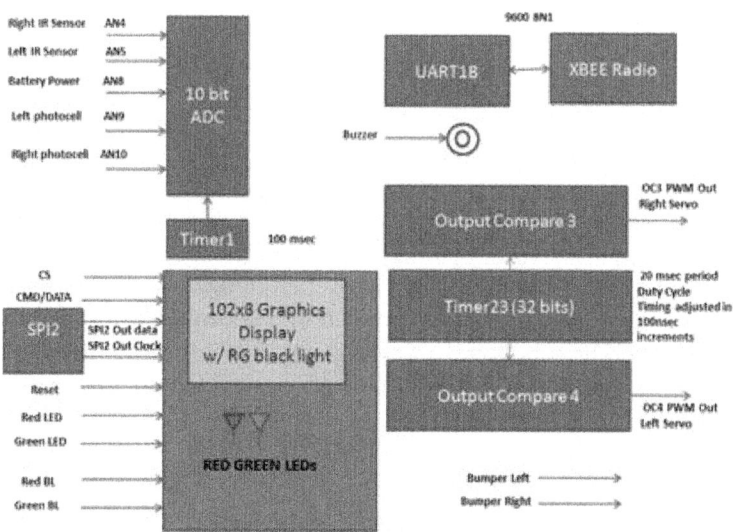

Fig. : Peripheral usage.

The PIC32 pinouts and their associated Experimenter connections.

PIC32 Peripheral Connections

- **ADC Converter Channels**
  - AN5 Left IR Sensor JP2 Pin 7
  - AN4 Right IR Sensor JP2 Pin 8
  - AN8 Battery Power JP4 Pin 8
  - AN9 Left Photo Cell BOT Pin 1
  - AN10 Right Photo Cell BOT Pin 2
- **PWM outputs**
  - OC3 Right Wheel Servo JP2 Pin 1
  - OC4 Left Wheel Servo JP2 Pin 2
- **Digital I/O**
  - RG9 Left Bumper In JP4 Pin 3
  - RG6 Right Bumper In JP4 Pin 5
  - RF1- Buzzer Out- TOP Pin 7
  - RD5 – XBEE Reset- BOT Pin 7

- **Uart1B**
  - U1BRX -XBEE DOUT-BOT Pin 5
  - U1BTX- XBEE DIN-BOT Pin 7
- **LCD Graphics Display**
  - RF4 Red LED JP3 Pin 5
  - Rd10 Green LED JP4 Pin 1
  - RB0 Chip Select JP3 Pin 1
  - RB3 Red LED Backlight JP3 Pin 4
  - RD4 Green LED Backlight JP4 Pin 2
  - RB1 Reset JP3 Pin 2
  - RB2 Command/Data JP3 Pin 3
  - SCK2A Clock JP4 Pin 6
  - SDO2A Data IN JP4 Pin 4

Fig. : PIC32 peripheral pinouts and EXP32 connectors.

The TOMBOT Motion Basics and the PIC32 Output Compare Peripheral

Let's review the basics for TOMBOT motor control. The servos use the Parallax (Futaba) Continuous Rotation Servos. With two-wheel control, the robot motion is controlled as per **Table below**.

Table : Robot motion.

Robot Motion	Left Servo	Right Servo
FORWARD	CCW	CW
BACKWARD	CW	CCW
LEFT TURN	CW	CW
RIGHT TURN	CCW	CCW
PIVOT LEFT	CENTER	CW
PIVOT RIGHT	CCW	CENTER
STOP	CENTER	CENTER

The servos are controlled by using a 20-ms (500-Hz) pulse PWM pattern where the PWM pulse can from 1.0 ms to 2.0 ms. The effects on the servos for the different PWM.

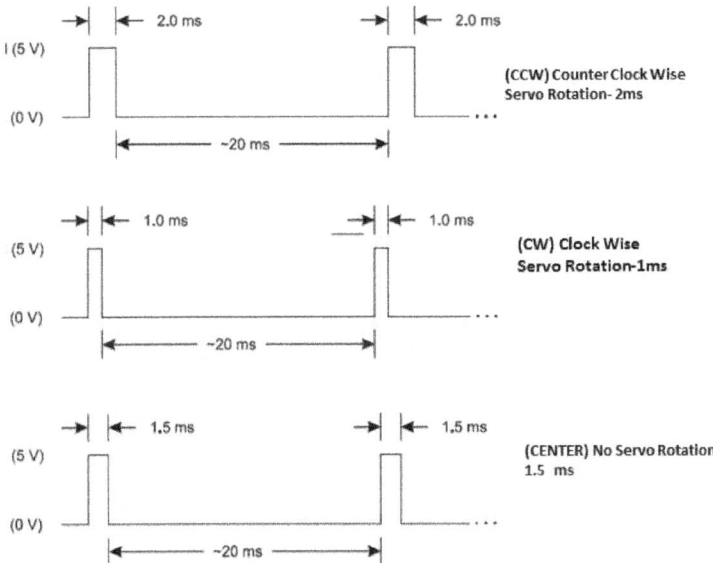

Fig. : Servo PWM control

The PIC32 microcontroller (used in the Experimenter) has five Output Compare modules (OCX, where X =1, 2, 3, 4, 5). We use two of these peripherals, specifically OC3, OC4 to generate the PWM to control the servo speed and direction. The OCX module can use either 16 Timer2 (TMR2) or 16 Timer3 (TMR3) or combined as 32-bit Timer23 as a time base and for period (PR) setting for the output pulse waveform. In our case, we are using Timer23 as a PR set to 20 ms (500 Hz). The OCXRS and OCXR registers are loaded with a 16-bit value to control width of the pulse generated during the output period. This value is compared against the Timer during each period cycle. The OCX output starts high and then when a match occurs OCX logic will generate a low on output. This will be repeated on a cycle-by-cycle basis.

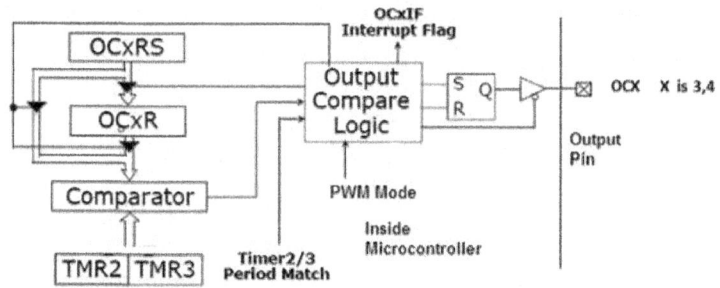

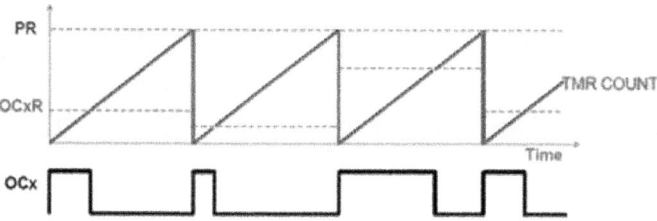

Fig. : PWM generation.

Chapter 3

ROBOTICS TECHNOLOGY, MECHANICS AND APPLICATIONS

Design, construction, and use of machines (robots) to perform tasks done traditionally by human beings. Robots are widely used in such industries as automobile manufacture to perform simple repetitive tasks, and in industries where work must be performed in environments hazardous to humans. Many aspects of robotics involve artificial intelligence; robots may be equipped with the equivalent of human senses such as vision, touch, and the ability to sense temperature.

Some are even capable of simple decision making, and current robotics research is geared toward devising robots with a degree of self-sufficiency that will permit mobility and decision-making in an unstructured environment. Today's industrial robots do not resemble human beings; a robot in human form is called an android. You have heard of the word "robots" during all your live; however you do not heard about the word "robotics" to often. Let star with a definition of the word itself; **Robotics** is a science of modern technology of general purpose of programmable machine systems. Contrary to the popular fiction image of robot as ambulatory machines of human appearance capable of performing almost any task. Most robotic systems are anchored to fixed positions in reality with limit mobility.

Robots perform a flexible, but restricted, number of operations in computer-aided manufacturing processes. These systems minimally contain a computer or a programmable device to control operations and effecters, devices that perform the desired work. The next paragraph represents the vision or general definition of robots according to the scientific knowledge and technology of that era.

ROBOTICS

Robotics is the branch of technology that deals with the design, construction, operation, and application of robots, as well as computer systems for their control, sensory feedback, and information processing. These technologies deal with automated machines that can take the place of humans in dangerous environments or manufacturing processes, or resemble humans in appearance, behavior, and/ or cognition. Many of today's robots are inspired by nature contributing to the field of bio-inspired robotics.

The concept of creating machines that can operate autonomously dates back to classical times, but research into the functionality and potential uses of robots did not grow substantially until the 20th century. Throughout history, robotics has been often seen to mimic human behavior, and often manage tasks in a similar fashion. Today, robotics is a rapidly growing field, as technological advances continue, research, design, and building new robots serve various practical purposes, whether domestically,commercially, or militarily. Many robots do jobs that are hazardous to people such as defusing bombs, mines and exploring shipwrecks.

Etymology

The word *robotics* was derived from the word *robot*, which was introduced to the public by Czech writer Karel Čapek in his play *R.U.R. (Rossum's Universal Robots)*, which was published in 1920. The word *robot* comes from the Slavic word *robota*, which means labour. The play begins in a factory that makes artificial people called *robots*, creatures who can be mistaken for humans – similar to the modern ideas of androids. Karel Čapek himself did not coin the word. He wrote a short letter in reference to an etymology in the *Oxford English Dictionary* in which he named his brother Josef Čapek as its actual originator.

According to the *Oxford English Dictionary*, the word *robotics* was first used in print by Isaac Asimov, in his science fiction short story "Liar!", published in May 1941 in*Astounding Science Fiction*. Asimov was unaware that he was coining the term; since the science and technology of electrical devices is *electronics*, he assumed *robotics*already referred to the science and technology of robots. In some of Asimov's other works, he states that the first use of the word *robotics* was in his short story *Runaround* (Astounding Science Fiction, March 1942). However, the original publication of "Liar!" predates that of "Runaround" by five months, so the former is generally cited as the word's origin.

History of Robotics

In 1927 the *Maschinenmensch* ("machine-human") gynoid humanoid robot (also called "Parody", "Futura", "Robotrix", or the "Maria impersonator") was the first depiction of a robot ever to appear on film was played by German actress Brigitte Helm in Fritz Lang's film Metropolis.

In 1942 the science fiction writer Isaac Asimov formulated his Three Laws of Robotics.

In 1948 Norbert Wiener formulated the principles of cybernetics, the basis of practical robotics.

Fully autonomous robots only appeared in the second half of the 20th century. The first digitally operated and programmable robot, the Unimate, was installed in 1961 to lift hot pieces of metal from a die casting machine and stack them. Commercial and industrial robots are widespread today and used to perform jobs more cheaply, or more accurately and reliably, than humans. They are also employed in jobs which are too dirty, dangerous, or dull to be suitable for humans. Robots are widely used inmanufacturing, assembly, packing and packaging, transport, earth and space exploration, surgery, weaponry, laboratory research, safety, and the mass production of consumer and industrial goods.

COMPONENTS

Power source

At present mostly (lead-acid) batteries are used as a power source. Many different types of batteries can be used as a power source for robots. They range from lead acid batteries which are safe and have relatively long shelf lives but are rather heavy to silver cadmium batteries that are much smaller in volume and are currently much more expensive. Designing a battery powered robot needs to take into account factors such as safety, cycle lifetime and weight.

Generators, often some type of internal combustion engine, can also be used. However, such designs are often mechanically complex and need fuel, require heat dissipation and are relatively heavy. A tether connecting the robot to a power supply would remove the power supply from the robot entirely. This has the advantage of saving weight and space by moving all power generation and storage components elsewhere. However, this design does come with the drawback of constantly having a cable connected to the robot, which can be difficult to manage. Potential power sources could be:

- pneumatic (compressed gases)
- hydraulics (liquids)
- flywheel energy storage
- organic garbage (through anaerobic digestion)
- faeces (human, animal); may be interesting in a military context as faeces of small combat groups may be reused for the energy requirements of the robot assistant

Actuation

Actuators are like the "muscles" of a robot, the parts which convert stored energy into movement. By far the most popular actuators are electric motors that spin a wheel or gear, and linear actuators that control industrial robots in factories. But there are some recent advances in alternative types of actuators, powered by electricity, chemicals, or compressed air.

Fig. : A robotic leg powered by air muscles.

Electric Motors

The majority of robots use electric motors, often brushed and brushless DC motors in portable robots, or AC motors in industrial robots and CNC machines. These motors are often preferred in systems with lighter loads, and where the predominant form of motion is rotational.

Linear Actuators

Various types of linear actuators move in and out instead of rotating, and often have quicker direction changes, particularly when very large forces are needed such as with industrial robotics. They are typically powered by compressed air (pneumatic actuator) or an oil (hydraulic actuator).

Series Elastic Actuators

A spring can be designed as part of the motor actuator, to allow improved force control. It has been used in various robots, particularly walking human-oid robots.

Air Muscles

Pneumatic artificial muscles, also known as air muscles, are special tubes that contract (typically up to 40%) when air is forced inside them. They have been used for some robot applications.

Muscle Wire

Muscle wire, also known as Shape Memory Alloy, Nitinol or Flexinol Wire, is a material that contracts slightly (typically under 5%) when electricity runs through it. They have been used for some small robot applications.

Electroactive Polymers

EAPs or EPAMs are a new plastic material that can contract substantially (up to 380% activation strain) from electricity, and have been used in facial muscles and arms of humanoid robots, and to allow new robots to float, fly, swim or walk.

Piezo Motors

Recent alternatives to DC motors are piezo motors or ultrasonic motors. These work on a fundamentally different principle, whereby tiny piezoceramic elements, vibrating many thousands of times per second, cause linear or rotary motion. There are different mechanisms of operation; one type uses the vibration of the piezo elements to walk the motor in a circle or a straight line. Another type uses the piezo elements to cause a nut to vibrate and drive a screw. The advantages of these motors are nanometerresolution, speed, and available force for their size. These motors are already available commercially, and being used on some robots.

Elastic Nanotubes

Elastic nanotubes are a promising artificial muscle technology in early-stage experimental development. The absence of defects in carbon nanotubes enables these filaments to deform elastically by several percent, with energy storage levels of perhaps $10 J/cm^3$ for metal nanotubes. Human biceps could be replaced with an 8 mm diameter wire of this material. Such compact "muscle" might allow future robots to outrun and outjump humans.

Sensing

Sensors allow robots to receive information about a certain measurement of the environment, or internal components. This is essential for robots to perform their tasks, and act upon any changes in the environment to calculate the appropriate response. They are used for various forms of measurements, to give the robots warnings about safety or malfunctions, and to provide real time information of the task it is performing.

Touch

Current robotic and prosthetic hands receive far less tactile information than the human hand. Recent research has developed a tactile sensor array that mimics the mechanical properties and touch receptors of human fingertips. The sensor array is constructed as a rigid core surrounded by conductive fluid contained by an elastomeric skin. Electrodes are mounted on the surface of the rigid core and are connected to an impedance-measuring device within the core. When the artificial skin touches an object the fluid path around the electrodes is deformed, producing impedance changes that map the forces received from the object. The researchers expect that an important function of such artificial fingertips will be adjusting robotic grip on held objects.

Scientists from several European countries and Israel developed a prosthetic hand in 2009, called SmartHand, which functions like a real one — allowing patients to write with it, type on a keyboard, play piano and perform other fine movements. The prosthesis has sensors which enable the patient to sense real feeling in its fingertips.

Vision

Computer vision is the science and technology of machines that see. As a scientific discipline, computer vision is concerned with the theory behind artificial systems that extract information from images. The image data can take many forms, such as video sequences and views from cameras.

In most practical computer vision applications, the computers are pre-programmed to solve a particular task, but methods based on learning are now becoming increasingly common.

Computer vision systems rely on image sensors which detect electromagnetic radiation which is typically in the form of either visible light or infra-red light. The sensors are designed using solid-state physics. The process by which light propagates and reflects off surfaces is explained using optics. Sophisticated image sensors even requirequantum mechanics to provide a complete understanding of the image formation process. Robots can also be equipped with multiple vision sensors to be better able to compute the sense of depth in the environment. Like human eyes, robots' "eyes" must also be able to focus on a particular area of interest, and also adjust to variations in light intensities.

There is a subfield within computer vision where artificial systems are designed to mimic the processing and behavior of biological system, at different levels of complexity. Also, some of the learning-based methods developed within computer vision have their background in biology.

Other

Other common forms of sensing in robotics use LIDAR, RADAR and SONAR.

Manipulation

Robots need to manipulate objects; pick up, modify, destroy, or otherwise have an effect. Thus the "hands" of a robot are often referred to as *end effectors*, while the "arm" is referred to as a *manipulator*. Most robot arms have replaceable effectors, each allowing them to perform some small range of tasks. Some have a fixed manipulator which cannot be replaced, while a few have one very general purpose manipulator, for example a humanoid hand.

Fig. : KUKA industrial robot operating in a foundry.

For the definitive guide to all forms of robot end-effectors, their design, and usage consult the book "Robot Grippers".

Mechanical Grippers

One of the most common effectors is the gripper. In its simplest manifestation it consists of just two fingers which can open and close to pick up and let go of a range of small objects. Fingers can for example be made of a chain with a metal wire run through it. Hands that resemble and work more like a human hand include the Shadow Hand, the Robonaut hand, ... Hands that are of a mid-level complexity include the Delft hand. An example of a simpler mechanical gripper is Cornell's universal jamming gripper, which does not use fingers but instead uses the principle of granular jamming to switch the gripper from deformable to solid.

Mechanical grippers can come in various types, including friction and encompassing jaws. Friction jaws use all the force of the gripper to hold the object in place using friction. Encompassing jaws cradle the object in place, using less friction.

Vacuum Grippers

Vacuum grippers are very simple astrictive devices, but can hold very large loads provided the prehension surface is smooth enough to ensure suction.

Pick and place robots for electronic components and for large objects like car screens, often use very simple vacuum grippers.

General Purpose Effectors

Some advanced robots are beginning to use fully humanoid hands, like the Shadow Hand, MANUS, and the Schunk hand. These are highly dexterous manipulators, with as many as 20 degrees of freedom and hundreds of tactile sensors.

LOCOMOTION

Rolling Robots

For simplicity most mobile robots have four wheels or a number of continuous tracks. Some researchers have tried to create more complex wheeled robots with only one or two wheels. These can have certain advantages such as greater efficiency and reduced parts, as well as allowing a robot to navigate in confined places that a four wheeled robot would not be able to.

Two-wheeled Balancing Robots

Balancing robots generally use a gyroscope to detect how much a robot is falling and then drive the wheels proportionally in the opposite direction, to counterbalance the fall at hundreds of times per second, based on the dynamics of an inverted pendulum. Many different balancing robots have been designed. While the Segway is not commonly thought of as a robot, it can be thought of as a component of a robot, when used as such Segway refer to them as RMP (Robotic Mobility Platform). An example of this use has been as NASA's Robonaut that has been mounted on a Segway.

One-wheeled Balancing Robots

A one-wheeled balancing robot is an extension of a two-wheeled balancing robot so that it can move in any 2D direction using a round ball as its only wheel. Several one-wheeled balancing robots have been designed recently, such as Carnegie Mellon University's "Ballbot" that is the approximate height and width of a person, and Tohoku Gakuin University's "BallIP". Because of the long, thin shape and ability to maneuver in tight spaces, they have the potential to function better than other robots in environments with people.

Spherical Orb Robots

Several attempts have been made in robots that are completely inside a spherical ball, either by spinning a weight inside the ball, or by rotating the outer shells of the sphere. These have also been referred to as an orb bot or a ball bot.

Six-wheeled Robots

Using six wheels instead of four wheels can give better traction or grip in outdoor terrain such as on rocky dirt or grass.

Tracked Robots

Tank tracks provide even more traction than a six-wheeled robot. Tracked wheels behave as if they were made of hundreds of wheels, therefore are very common for outdoor and military robots, where the robot must drive on very rough terrain. However, they are difficult to use indoors such as on carpets and smooth floors. Examples include NASA's Urban Robot "Urbie".

Walking Applied to Robots

Walking is a difficult and dynamic problem to solve. Several robots have been made which can walk reliably on two legs, however none have yet been made which are as robust as a human. There has been much study on human inspired walking, such as AMBER lab which was established in 2008 by the Mechanical Engineering Department at Texas A&M University. Many other robots have been built that walk on more than two legs, due to these robots being significantly easier to construct. Walking robots can be used for uneven terrains, which would provide better mobility and energy efficiency than other locomotion methods. Hybrids too have been proposed in movies such as I, Robot, where they walk on 2 legs and switch to 4 (arms+legs) when going to a sprint. Typically, robots on 2 legs can walk well on flat floors and can occasionally walk up stairs. None can walk over rocky, uneven terrain. Some of the methods which have been tried are:

ZMP Technique

The Zero Moment Point (ZMP) is the algorithm used by robots such as Honda's ASIMO. The robot's onboard computer tries to keep the total inertial forces (the combination of Earth's gravity and the acceleration and deceleration of walking), exactly opposed by the floor reaction force (the force of the floor pushing back on the robot's foot). In this way, the two forces cancel out, leaving no moment (force causing the robot to rotate and fall over). However, this is not exactly how a human walks, and the difference is obvious to human observers, some of whom have pointed out that ASIMO walks as if it needs the lavatory. ASIMO's walking algorithm is not static, and some dynamic balancing is used. However, it still requires a smooth surface to walk on.

Hopping

Several robots, built in the 1980s by Marc Raibert at the MIT Leg Laboratory, successfully demonstrated very dynamic walking. Initially, a robot with only one leg, and a very small foot, could stay upright simply by hopping. The movement is the same as that of a person on a pogo stick. As the robot falls to one side, it would jump slightly in that direction, in order to catch itself. Soon, the algorithm was generalised to two and four legs. A bipedal robot was demonstrated running and even performing somersaults. A quadruped was also demonstrated which could trot, run, pace, and bound.

Dynamic Balancing Controlled Falling

A more advanced way for a robot to walk is by using a dynamic balancing algorithm, which is potentially more robust than the Zero Moment Point technique, as it constantly monitors the robot's motion, and places the feet in order to maintain stability. This technique was recently demonstrated by Anybots' Dexter Robot, which is so stable, it can even jump. Another example is the TU Delft Flame.

Passive Dynamics

Perhaps the most promising approach utilizes passive dynamics where the momentum of swinging limbs is used for greater efficiency. It has been shown that totally unpowered humanoid mechanisms can walk down a gentle slope, using only gravity to propel themselves. Using this technique, a robot need only supply a small amount of motor power to walk along a flat surface or a little more to walk up a hill. This technique promises to make walking robots at least ten times more efficient than ZMP walkers, like ASIMO.

OTHER METHODS OF LOCOMOTION

Flying

A modern passenger airliner is essentially a flying robot, with two humans to manage it. The autopilot can control the plane for each stage of the journey, including takeoff, normal flight, and even landing. Other flying robots are uninhabited, and are known as unmanned aerial vehicles (UAVs). They can be smaller and lighter without a human pilot on board, and fly into dangerous territory for military surveillance missions. Some can even fire on targets under command. UAVs are also being developed which can fire on targets automatically, without the need for a command from a human. Other flying robots include cruise missiles, theEntomopter, and the Epson micro helicopter robot. Robots such as the Air Penguin, Air Ray, and Air Jelly have lighter-than-air bodies, propelled by paddles, and guided by sonar.

Snaking

Several snake robots have been successfully developed. Mimicking the way real snakes move, these robots can navigate very confined spaces, meaning they may one day be used to search for people trapped in collapsed buildings. The Japanese ACM-R5 snake robot can even navigate both on land and in water.

Skating

A small number of skating robots have been developed, one of which is a multi-mode walking and skating device. It has four legs, with unpowered wheels, which can either step or roll. Another robot, Plen, can use a miniature skateboard or roller-skates, and skate across a desktop.

Climbing

Several different approaches have been used to develop robots that have the ability to climb vertical surfaces. One approach mimics the movements of a human climber on a wall with protrusions; adjusting the center of mass and moving each limb in turn to gain leverage. An example of this is Capuchin, built by Dr. Ruixiang Zhang at Stanford University, California. Another approach uses the specialized toe pad method of wall-climbing geckoes, which can run on smooth surfaces such as vertical glass. Examples of this approach include Wallbot and Stickybot. China's "Technology Daily" November 15, 2008 reported New Concept Aircraft (ZHUHAI) Co., Ltd. Dr. Li Hiu Yeung and his research group have recently successfully developed the bionic gecko robot "Speedy Freelander".

According to Dr. Li introduction, this gecko robot can rapidly climbing up and down in a variety of building walls, ground and vertical wall fissure or walking upside down on the ceiling, it is able to adapt on smooth glass, rough or sticky dust walls as well as the various surface of metallic materials and also can automatically identify obstacles, circumvent the bypass and flexible and realistic movements. Its flexibility and speed are comparable to the natural gecko. A third approach is to mimic the motion of a snake climbing a pole.

Swimming (Piscine)

It is calculated that when swimming some fish can achieve a propulsive efficiency greater than 90%. Furthermore, they can accelerate and maneuver far better than any man-made boat or submarine, and produce less noise and water disturbance. Therefore, many researchers studying underwater robots would like to copy this type of locomotion. Notable examples are the Essex University Computer Science Robotic Fish, and the Robot Tuna built by the Institute of Field Robotics, to analyze and mathematically model thunniform motion. The Aqua Penguin, designed and built by Festo of Germany, copies the streamlined shape and propulsion by front "flippers" of penguins. Festo have also built the Aqua Ray and Aqua Jelly, which emulate the locomotion of manta ray, and jellyfish, respectively.

Sailing

Sailboat robots have also been developed in order to make measurements at the surface of the ocean. A typical sailboat robot is *Vaimos* built by IFREMER and ENSTA-Bretagne. Since the propulsion of sailboat robots uses the wind, the energy of the batteries is only used for the computer, for the communication and for the actuators (to tune the rudder and the sail). If the robot is equipped with solar panel, the robot could theoretically navigate forever. The two main competitions of sailboat robots are WRSCwhich takes place every year in Europe and Sailbot.

Environmental Interaction and Navigation

Though a significant percentage of robots in commission today are either human controlled, or operate in a static environment, there is an increasing interest in robots that can operate autonomously in a dynamic environment. These robots require some combination of navigation hardware and software in order to traverse their environment. In particular unforeseen events (*e.g.* people and other obstacles that are not stationary) can cause problems or collisions. Some highly advanced robots such as ASIMO, and Meinü robothave particularly good robot navigation hardware and software.

Also, self-controlled cars, Ernst Dickmanns' driverless car, and the entries in the DARPA Grand Challenge, are capable of sensing the environment well and subsequently making navigational decisions based on this information. Most of these robots employ a GPS navigation device with waypoints, along with radar, sometimes combined with other sensory data such as LIDAR, video cameras, and inertial guidance systems for better navigation between waypoints.

Human-robot Interaction

If robots are to work effectively in homes and other non-industrial environments, the way they are instructed to perform their jobs, and especially how they will be told to stop will be of critical importance. The people who interact with them may have little or no training in robotics, and so any interface will need to be extremely intuitive. Science fiction authors also typically assume that robots will eventually be capable of communicating with humans through speech, gestures, and facial expressions, rather than a command-line interface. Although speech would be the most natural way for the human to communicate, it is unnatural for the robot. It will probably be a long time before robots interact as naturally as the fictional C-3PO.

Speech Recognition

Interpreting the continuous flow of sounds coming from a human, in real time, is a difficult task for a computer, mostly because of the great variability of speech. The same word, spoken by the same person may sound different depending on local acoustics, volume, the previous word, whether or not the speaker has a cold, *etc.*. It becomes even harder when the speaker has a different accent. Nevertheless, great strides have been made in the field since Davis, Biddulph, and Balashek designed the first "voice input system" which recognized "ten digits spoken by a single user with 100% accuracy" in 1952. Currently, the best systems can recognize continuous, natural speech, up to 160 words per minute, with an accuracy of 95%.

Robotic Voice

Other hurdles exist when allowing the robot to use voice for interacting with humans. For social reasons, synthetic voice proves suboptimal as a communication medium,making it necessary to develop the emotional component of robotic voice through various techniques.

Gestures

One can imagine, in the future, explaining to a robot chef how to make a pastry, or asking directions from a robot police officer. In both of these cases, making hand gestureswould aid the verbal descriptions. In the first case, the robot would be recognizing gestures made by the human, and perhaps repeating them for confirmation. In the second case, the robot police officer would gesture to indicate "down the road, then turn right". It is likely that gestures will make up a part of the interaction between humans and robots.

Facial Expression

Facial expressions can provide rapid feedback on the progress of a dialog between two humans, and soon may be able to do the same for humans and robots. Robotic faces have been constructed by Hanson Robotics using their elastic polymer called Frubber, allowing a large number of facial expressions due to the elasticity of the rubber facial coating and embedded subsurface motors (servos). The coating and servos arc built on a metal skull. A robot should know how to approach a human, judging by their facial expression and body language. Whether the person is happy, frightened, or crazy-looking affects the type of interaction expected of the robot. Likewise, robots likeKismet and the more recent addition, Nexi can produce a range of facial expressions, allowing it to have meaningful social exchanges with humans.

Artificial Emotions

Artificial emotions can also be generated, composed of a sequence of facial expressions and/or gestures. As can be seen from the movie Final Fantasy: The Spirits Within, the programming of these artificial emotions is complex and requires a large amount of human observation. To simplify this programming in the movie, presets were created together with a special software program. This decreased the amount of time needed to make the film. These presets could possibly be transferred for use in real-life robots.

Personality

Many of the robots of science fiction have a personality, something which may or may not be desirable in the commercial robots of the future. Nevertheless, researchers are trying to create robots which appear to have a personality: *i.e.* they use sounds, facial expressions, and body language to try to convey an internal state, which may be joy, sadness, or fear. One commercial example is Pleo, a toy robot dinosaur, which can exhibit several apparent emotions.

CONTROL OF ROBOT

The mechanical structure of a robot must be controlled to perform tasks. The control of a robot involves three distinct phases – perception, processing, and action (robotic paradigms). Sensors give information about the environment or the robot itself (*e.g.* the position of its joints or its end effector). This information is then processed to calculate the appropriate signals to the actuators (motors) which move the mechanical.

The processing phase can range in complexity. At a reactive level, it may translate raw sensor information directly into actuator commands. Sensor fusion may first be used to estimate parameters of interest (*e.g.* the position of the robot's gripper) from noisy sensor data. An immediate task (such as moving the gripper in a certain direction) is inferred from these estimates. Techniques from control theory convert the task into commands that drive the actuators.

At longer time scales or with more sophisticated tasks, the robot may need to build and reason with a "cognitive" model. Cognitive models try to represent the robot, the world, and how they interact. Pattern recognition and computer vision can be used to track objects. Mapping techniques can be used to build maps of the world. Finally, motion planning and other artificial intelligencetechniques may be used to figure out how to act. For example, a planner may figure out how to achieve a task without hitting obstacles, falling over, *etc.*

Autonomy levels

Control systems may also have varying levels of autonomy.

1. Direct interaction is used for haptic or tele-operated devices, and the human has nearly complete control over the robot's motion.
2. Operator-assist modes have the operator commanding medium-to-high-level tasks, with the robot automatically figuring out how to achieve them.
3. An autonomous robot may go for extended periods of time without human interaction. Higher levels of autonomy do not necessarily require more complex cognitive capabilities. For example, robots in assembly plants are completely autonomous, but operate in a fixed pattern.

Another classification takes into account the interaction between human control and the machine motions.

1. **Teleoperation.** A human controls each movement, each machine actuator change is specified by the operator.
2. **Supervisory.** A human specifies general moves or position changes and the machine decides specific movements of its actuators.
3. **Task-level autonomy.** The operator specifies only the task and the robot manages itself to complete it.
4. **Full autonomy.** The machine will create and complete all its tasks without human interaction.

GENERAL DEFINITION FOR ROBOT

"A re-programmable, multifunctional mechanical manipulator designed to move material, parts, tools, or specialized devices through various programmed motions for the performance of a variety of tasks."

— *From the Robot Institute of America, 1979*

Importance of Robotics

The response is simple **Robotics** is a science that combines a range of fields like Mechanical Engineering, Electrical Engineering, and Computer Science. Robotics is ideal for adolescent students because it exposes them to hands-on applications of math, science, and engineering concepts. In addition, robotics motivates potential scientists and engineers to understand how things work and encourages them to use their imagination to create new technologies and improve old technologies. The next part of this background should cover the main components of a robot including some basic concepts for third to fifth grade.

Now a day thinks are getting sophisticated with more technological advance. A new perception and vision of the robot representation includes the following characteristics:

Robot Components

- **Mechanical platforms — or hardware base** is a mechanical device, such as a wheeled platform, arm, fixed frame or other construction, capable of interacting with its environment and any other mechanism involve with his capabilities and uses.

- **Sensors systems** is a special feature that rest on or around the robot. This device would be able to provide judgment to the controller with relevant information about the environment and give useful feedback to the robot.

- **Joints** provide more versatility to the robot itself and are not just a point that connects two links or parts that can flex, rotate, revolve and translate. Joints play a very crucial role in the ability of the robot to move in different directions providing more degree of freedom.

- The **controller** functions as the "brain" of the robot. Robots today have controllers that are run by programs - sets of instructions written in code. In other words, it is a computer used to command the robot memory and logic. So it, be able to work independently and automatically.

- **Power Source** is the main source of energy to fulfill all the robots needs. It could be a source of direct current as a battery, or alternate current from a power plant, solar energy, hydraulics or gas.

- **Artificial intelligence** represents the ability of computers to "think" in ways similar to human beings. Present day "AI" does allow machines to mimic certain simple human thought processes, but can not begin to match the quickness and complexity of the brain. On the other hand, not all robots possess this type of capability. It requires a lot of programming and sophisticates controllers and sensorial ability of the robot to reach this level.

The simplest behavior of a robot is **locomotion**. Typically, joints and wheels are used as the underlying mechanism to make a robot move from one point to the next.

This type of motion should include the **adaptability** and **versatility** of the robot to continue with a specific task. **Adaptability** means adjustment to the task being carried out. **Versatility** means that the robot should have such a mechanical structure that it can carry out different tasks or perhaps the same task in different ways.

Beginning of Robot

The word "robot" has its origin from the German word "robat". It appears that the science fiction writer Isaac Asimov was the first to use the word "robotics" to describe robot technology.

The first robots

Joseph Engel Berger, in the picture, is entitled to be the father of robotics, together with George Deroe developed the first commercial robot, **Unimate**, in 1961. It was placed on Ford and was there used for a press-loading operation.

Unimate

The first robots were principally intended to replacing humans in monotonous, heavy and hazardous processes. Distinctive features of the use of the newly developed robots were in handling of materials and work pieces without direct control or participation in the manufacturing process. Robots did not become a major force in industry generally until they had been used extensively in the Japanese automobile industry.

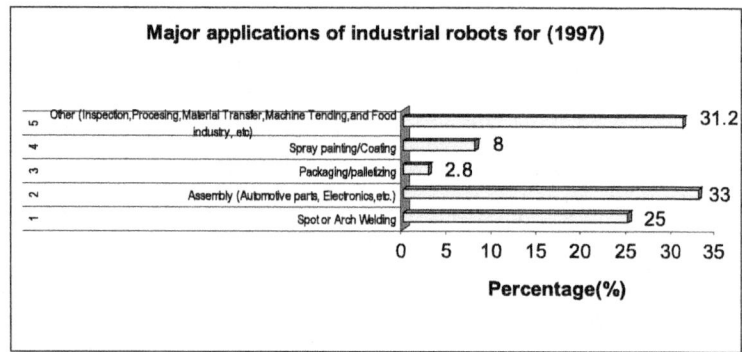

Fig. : : Shows the percentage of applications of robots at the industry during 1997.

Physical Robot Configurations

Over the years robot manufacturers have developed many types of robots of differing configurations and mechanical design, to give a variety of spatial arrangements and working volumes. These have evolved into six common types of system:

Table : Physical Configurations.

Cartesian robot it is form by 3 prismatic joints, whose axes are coincident with the X, Y and Z planes. These robots move in three directions, in translation, at right angles to each other.	**SCARA robot** which stands for Selective Compliance Assembly Robot Arm it is built with 2 parallel rotary joints to provide compliance in a plane. The robots work in the XY-plane and have Z-movement and a rotation of the gripper for assembly.
Cylindrical robot is able to rotate along his main axes forming a cylindrical shape. The robot arm is attached to the slide so that it can be moved radially with respect to the column.	**Articulated robots** are mechanic manipulator that looks like an arm with at least three rotary joints. They are used in welding and painting; gantry and conveyor systems move parts in factories.
Spherical robot is able to rotate in two different directions along his main axes and the third joint moves in translation forming a hemisphere or polar coordinate system. It used for a small number of vertical actions and is adequate for loading and unloading of a punch	**Parallel robot** is a complex mechanism which is constituted by two or more kinematics chains between, the base and the platform where the end-effectors are located. Good examples are the flying simulator and 4-D attractions at Univ. Studios

"**Workspace envelope**" is one of the new terms that are going to be covered in the following table. It really describes how the robot is constrained by its mechanical systems configuration. Each joint of a robot has a limit of motion range. A workspace envelope of a robot is defined as all the points in the surrounding space that can be reached by the robot. Clear under standing of the workspace envelope of a robot to be used is important because all interaction with other machines, parts, and processes only takes place within this volume of space.

ROBOTICS TECHNOLOGY TODAY

Ever faster processors, cheaper sensors, abundant open-source code, ubiquitous connectivity, and the advent of 3D printing are some of the forces behind the recent proliferation of robots.

Some observers are voicing their fears about a decline in human-human interaction, while others warn of an irreversible and senseless loss of jobs, with robots taking over tasks that, they argue, should not be performed by machines (such as caring for the elderly). Trade-offs will certainly be part of our growing reliance on robotics and automation. And it will be up to us to manage these trade-offs, just as we have with other technologies such as electricity, the automobile, aviation, nuclear power, computers, and the Internet.

Fig. : Tesla Motors.

As a VC looking for investment opportunities in robotics, I talk to lots of different people about their views on the future of this industry. Many times what I hear from these people are totally contradictory, so I often have to come up with my own conclusions.

Robots are Intended To Eliminate Jobs: MYTH

Almost every major manufacturing and logistics company I've spoken to looks to robotics as a means to improve the efficiency of its operations and the quality of life of its existing workers. So human workers continue to be a key part of the business when it comes to robotics. In fact, workers should view robots as how skilled craftsmen view their precision tools: enhancing output while creating greater job satisfaction. Tesla Motors is just one example of using robots to do all the limb-threatening and back-breaking tasks while workers oversee their operation and ensure the quality of their output. At Tesla's assembly lines, robots glue, rivet, and weld parts together, under the watchful eye of humans. These workers can pride themselves with being part of a new era in manufacturing where robots help to reinvent and reinvigorate existing industries like the automotive sector.

Manufacturing and Logistics Must Adopt Robots to Survive: FACT

Although total cost of ownership is a popular yardstick used for purchasing capital equipment, payback time is more commonly used for automating basic (typically arduous) worker tasks. If we use pick-and-place as an example, one estimate says that each work "cell" (or equivalent of one worker) costs $32.5k/ year (not including benefits) for a single shift in the United States. Historically, manufacturing and logistics companies have adopted automation equipment with one-year payback. Therefore, a $65k installed robot, operating on two shifts, is highly desirable. However, the net present value of down time, upgrades, and maintenance must be calculated into the $65k number. Labor costs are lower in

Asia, where each work "cell" is closer to $20k, though that number is expected to increase sharply.

It's not surprising, then, that Asian manufacturers, including Flextronics and Foxconn, are actively seeking automation technology to increase productivity of their existing workforce and meet increasing demand. And we are already seeing a similar trend in the logistics sector. Deregulation and technology brought about the notion of the third-party logistics provider, also known as the 3PL, which offers transportation, warehousing, and pick-and-place to suppliers in a wide variety of product categories and marketplaces. 3PLs compete on their ability to deliver higher-quality services as lower cost to their clients. With technology being responsible for the 3PL's coming into existence, they continue to spend vast sums on automation in each of their hundreds of warehouses to improve the quality and variety of their logistics offerings — in other words, robotics will allow them to remain competitive and survive.

Autonomous Robots are Still Too Slow: FACT

If you were watching the DARPA Robotics Challenge last year, you would be quick to notice that the robots are awkward and slow at completing even the most basic tasks. Although Moore's Law has sped up capabilities like machine vision, traditional search-based algorithms for navigating through decision trees are just too slow in runtime. As the famous video of PR2 folding towelsshowed, tasks that we take for granted as humans are time-intensive for algorithms — even with the help of the most advanced sensors — to navigate. Significant advances can be made with fundamental algorithmic improvements that take the "brute force" out of machine vision and other tasks. In particular, lots of research are going into pattern recognition, in addition to handing control over to the "cloud" when machine vision algorithms come short.

Robots are Too Expensive: MYTH

Modern household appliances are examples of dedicated pieces of hardware that people buy without thinking twice. These devices benefit from decades of incremental improvement, and millions of units in the field over which development and tooling costs are spread out. The same could apply to robots. The problem, though, is that robots still require specialized — and costly — hardware. In particular, actuators are among the most expensive parts in almost every robot, and unlike processors and sensors like cameras, the cost of actuators is not coming down at a significant pace. Consider, for example, Willow Garage spinout Industrial Perception. Before being acquired by Google, the startup managed to get close to human speed in identifying and unloading assorted boxes from a container.

But although its robotic system used inexpensive sensors like a few cameras and Microsoft Kinect devices, it still required a relatively expensive robot arm powered by conventional actuators. The good news is I believe we're about to see a lot of innovation in actuation systems — and that will finally bring the cost of robots down, as it happened to appliances. One promising area involves the special sensor-equipped joints that allow robots to control their motions in a precise and safe way. Groups such as CMU-spinout IAM Robotics, Redwood Robotics(acquired by Google), and Modbot are using unique approaches to reduce the number and simplify the motors, gears, and sensors needed, hence dramatically reducing the cost of the arms, which in most cases dominate the cost of the entire robot.

Robots are Difficult to Use: FACT

Rethink Robotics' Baxter is one example of a robot designed to be affordable and simple to program. However, those qualities come at the expense of the speed and precision associated with traditional industrial robots. Most autonomous robots require highly-trained workers and painstaking programming, calibrating, and testing. These requirements are often unacceptable for businesses that expect capital equipment to be 100 percent operational within days, if not hours of shipment, without hiccups. Although efforts like the Robot Operating System (ROS) and Open Source Computer Vision (OpenCV) are trying to simplify tasks and requirements to have robots up and running and doing useful things, they're still mostly used by experienced roboticists with PhD degrees. Would the personal computer have been as popular as it is today if they booted to a command prompt? They probably would amongst developers, but not among the masses that drove the PC, and later the Internet, revolutions. Standard CAD/CAM/Gerber/toolpath generators made automated machine tools a no-brainer. We need the equivalent for robotics.

SIMPLE ROBOTICS MECHANICS

Machine

Is a device that transmits, or changes the application of energy to carry out a quantity of work. It allows the multiplication of force at the expense of distance. Work is defined as a force applied through a distance.

Simple Machines

Simple machines have existed and have been used for centuries. Each machine makes work easier to do. Each of them provides some trade-off between the force applied and the distance over which the force is applied.

Levers

A **lever** is a stiff bar that rotates about a pivot point called the fulcrum. The lever consists of three parts. The fulcrum, load (it acts on the rod) and a rod (holds the load or applied effort). Levers are classified into three classes. Depending on where the pivot point is located, a lever can multiply either the force applied or the distance over which the force is applied.

Levers are classified into three classes:

1. **First Class Levers** that has a turning point between the apply force and the load. A seesaw is an example of a simple first class lever. A pair of scissors is an example of two connected first class levers.

2. **Second Class Levers** has his load between the pivot and the apply force. A wheelbarrow is an example of a simple second class lever. A nutcracker is an example of two connected second class levers.

3. Third Class Levers the effort is between the pivot and the load. A stapler or a fishing rod is an example of a simple third class lever. A pair of tweezers is an example of two connected third class levers.

Gears

Gears and chains are mechanical platforms that provide a strong and accurate way to transmit rotary motion from one place to another, possibly changing it along the way. The speed change between two gears depends upon the number of teeth on each gear. When a powered gear goes through a full rotation, it pulls the chain by the number of teeth on that gear.

When you have two gears of different size in movement the smaller gear spins twice as fast as the larger gear because the diameter of the gear on the right is twice that of the gear on the left. The gear ratio is therefore 2:1 pronounced, ("Two to one"). This gear ratio is directly proportional with the amount of torque. In a simple way the gear that spins twice as fast generates the lowest torque.

Gears are generally used for one of four different reasons:

1. Reverse the direction of rotation.
2. Increase or decrease the speed of rotation or torque.
3. Shift a rotational motion to a different axis.
4. To keep the rotation of two axes synchronized.

Pulleys

Pulleys and belts are two types of mechanical platforms used in robots, work the same way as gears and chains. These kinds of pulleys are wheels with a groove around the edge, and belts are the rubber loops that fit in that groove.

In addition to the pulley describe on the previous paragraph they are other types of **pulleys** that are made up of a rope or chain and a wheel around which fits the rope. When you pull down one end of the rope goes up.

There are two types of pulley and both are classified by its movement.

The first type is a **fixed pulley** that is attached permanently to a surface or place. This type of pulley uses more effort to lift the load from the ground. The second type is a **movable pulley** that is free to travel along the rope or chain path following the load direction. The movable pulley allows the effort to be less than the load weight. A third type could be defined as a combined **pulley.** It diminishes the effort needed to lift huge loads dropping this effort in less than half of the load weight.

Gearbox

It operates on the same principles as the gear and chain, but without the chain. Gearboxes require closer tolerances, since instead of using a large loose chain to transfer force and adjust for misalignments, the gears mesh directly with each other. Examples of gearboxes can be found on the transmission in a car, the timing mechanism in a grandfather clock, and the paper-feed of your printer.

Electric Motor

An **Electric Motor** is a machine which converts electric energy into mechanical energy.

When an electric current is passed through a wire loop that is in a magnetic field, the loop will rotate and the rotating motion is transmitted to a shaft, providing useful mechanical work. The traditional electric motor consists of a conducting loop that is mounted on a rotational shaft. The electrical current fed in by carbon blocks, called brushes, and enters the loop through two slip rings. The magnetic field around the loop, supplied by an iron core field magnet, causes the loop to turn when current is flowing through it.

A variety of electric motors provide power to robots, allowing them to move material, parts, tools, or specialized devices with various programmed motions. The **efficiency** of a motor describes how much of the electrical energy utilize is converted to mechanical energy.

The difference between Direct Current (DC) and Alternating Current (AC) electric current is the way that electrons travel in the wire connections.

1. **Alternating Current (AC):** is the type of electricity that we get from plugs in the wall. In an alternating current all of the electric charges switch their direction of flow back and forth.

2. **Direct current (DC):** is the continuous flow of electricity through a conductor such as a wire from high to low potential. The direct current electric charges flow always in the same direction.

DIFFERENT TYPES OF MOTORS

Direct Current (DC) Motor

In this motor a device known as a **split ring commutator** switches the direction of the electric current at each half of the rotation of the rotor. This is due to keep the shaft motion direction unchanged. In any motor the stationary parts constitute the **stator**, and the assembly carrying the loops is called the **rotor**, or

armature. As it is easy to control the speed of direct-current motors by varying the field or armature voltage, these are used where speed control is necessary.

Brushless DC Motors

This kind of motor is constructed in a reverse fashion from the traditional form. The rotor contains a permanent magnet and the stator has the conducting coil of wire. By the elimination of brushes, this motor reduced maintenance, no spark hazard, and better speed control. They are widely used in computer disk drives, tape recorders, and other electronic devices.

Alternating Current (AC) motor

This kind of motor works with the electrical current flow in the laminate core loop. The electrical current is synchronized to reverse direction when the laminate core loop plane is perpendicular to the magnetic field and there is no magnetic force exerted on the loop. In alternating current induction motors the current passing through the loop does not come from an external source but is induced as the laminate core passes through the magnetic field. The speed of AC induction motors is set roughly by the motor construction and the frequency of the current. To control the motor speed it's necessary to use a mechanical transmission. To obtain greater flexibility, the rotor circuit can be connected to various external control circuits.

Synchronous AC Motors

This motor is designed to operate exclusively on alternating current and is essentially identical to the generator. A **generator** uses work to produce electric energy while a **motor** uses electric energy to produce **work**. If you connect a synchronous AC motor to the power line and let it turn, it will draw energy out of the electric circuit and provide work. But if you connect that same motor to a light bulb and turn its rotor by hand, it will generate electricity and light the bulb. In addition, the more work the motor does, the more electric energy it consumes.

How this Motor Works?

The rotor is a permanent magnet that spins between two stationary electromagnets. Each time the rotor's is about to reach stationary electromagnet, the current reverses. This cycle maintain the rotor mechanism turns endlessly.

Because its rotation is perfectly synchronized with the current reversals, this motor is called a synchronous AC electric motor. When a **Synchronous AC Motor's coils** become hot when large currents flow through them. Whether a motor is consuming or producing electric power, it will overheat and burn out when it handle too much current. Failures of this type occur in overloaded motors and power plant generators during periods of exceptionally high electric power usage. Circuit breakers are often used to stop the current flow before it can cause damage

Universal Motors

This intermediate motor works on either direct or alternate electric current. In fact a DC motor can not tolerate alternate current. A real AC motor can not tolerate direct electric current because it depends on the electrical line's to reverse the current direction flow going back and forth and keeps the rotor moving.

However, if you replace the permanent magnets of a DC motor with electromagnets and connect these electromagnets in the same circuit as the commutator and rotor, you will have a universal motor. This motor will spin properly when powered by either direct or alternating current. If you connect direct current to a universal motor, the stationary electromagnets will behave as if they were permanent magnets and the universal motor will operate just like a DC motor. Since the universal motor always turns in the same direction, regardless of which way current flows through it, it will works just fine with alternate current power.

A simple motor has eight parts, as shown in the diagram below:

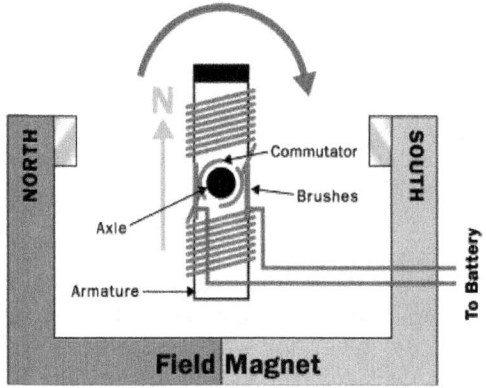

Armature or rotor: is a set of electromagnets. This structure supports the conductors that cut the magnetic field and carry the exciting electric current in a motor.

Commutator: a series of bars or segments so connected to armature coils of a generator or motor that rotation of the armature will in conjunction with fixed brushes result in unidirectional current in the reversal of the current into the coils in the case of a motor.

Brushes: are the lifelines of the motor and allows the electric current to flow into the rotor once it touches one of these plates and leaves the rotor through a second brush that touches the other plate use. They get worn and burnt.

Axle or drive shaft: Is the mechanism in charge of transmitting the torque from the motor to any other mechanism that requires power to realize a work.

Electric Coil: is a set of Cooper windings that goes around the armature it provides the pathway for the electric current around the DC motor.

Cooper winding is characterize by a single wire use to build the electric coils use on a motor.

Field magnet: is a magnet for producing and maintaining a magnetic field especially in an electric motor.

Power supply: of some sort DC (direct current) source such as a battery, and motors which are powered by an AC (alternating current) source.

Table : Enumerate the basic Direct Current Power Supplies uses in robotics.

Size	NEDA	IEC	Description
AAA	24A	LR03	Smallest of the command sizes
AA	15A	LR6	Most popular small battery, typically used in packs of 2 or 4
C	14A	LR14	Small flashlight battery, large toys
D	13A	LR20	Largest common battery
9v	1604A	6L-R61	Rectangular with clip-on connector

Electric Circuits Schematics

Students should be aware of the importance of an electric circuit, especially in their everyday life. When we connect various components together with electric wires, we create an electric circuit. The electrons must have a voltage source that is supply by a Power Source (Battery, Alternator, Generator, *etc.*) to create their movement. The electrons path configuration is responsible for the way that circuits are name nowadays. There are two main types of current electric circuits, series and parallel. A third type can be obtained as a combination of the two basic type of circuit and it can be name as a **series-parallel circuit.**

A simple **series circuit** is attached; to a single pathway where the electric current will flow. In a series circuit, when one of the bulbs or one of the wires is left open or is broken, the entire circuit breaks. A parallel circuit is designed so if one branch is defective, the flow of electricity will not be broken to the other branches. These individual branches keep the flow of electrons for different circuit components. Both series and parallel connection have their own distinctive characteristics. A **series-parallel circuit** is more often use in building, houses and other commercial structures. It combines the characteristics of the first two types of circuits.

They are three different circuit types; Series Circuit, Parallel Circuit, and Series-Parallel Circuit all require the same basic components:

1. Power Source (Battery, Alternator, Generator, *etc.*)
2. Protection Device (Fuse, Fusible Link, or Circuit Breaker)
3. Load Device (Lamp, Motor, Winding, Resistor, *etc.*
4. Control (Switch, Relay, or Transistor)
5. Conductors (A Return Path, Wiring to Ground)

Circuit Symbols

Circuit symbols are used in circuit diagrams to show how a circuit is connected together electrically. They are used for designing and testing circuits, and understand how they work. To build a circuit you need a diagram that shows the layout of the components on printed circuit board. A circuit a board is the one that takes care of all the individual components.

Advanced Robotics Technology

Robot technology is not only entering new industries but it is also entering the domestic field. Companies have been working on home security automation in which robot technology watches for invaders. If an invader is caught it lets out a scream while calling the police. They are also working on robotic technology that can give business tours to executives with out having them present *via* cameras and microphones. There may also be robots that will be able to remind you to take your medications. And you may soon find yourself with some extra hands around the house with robots that can wash windows, vacuum, and even mow the lawn.Robots are also making their way into the medical field by assisting in high precision sugeries such as gall bladder removals, brian surgeries, and heart bypasses. While robotic cameras are being used in some settings for diagnostics. Pharmaceuticals are also jumping on the robotics band wagon to use robots for testing quality control by food processors to check for bacterial comtamination. Causes for Increased Robotic TechnologyIt has been more than just advances in technology for the increase in robotic use, as industrial robot sales are expected to increase by $3 billion a year through 2015. It is also due to 2 major industries, the green tech industry and the military. The green tech industry uses advanced robotics technology for solar cells, wind turbines, and auto batteries for plug-in vehicles.The military is one of the largest users of robotics. It has been using robots for battlefields and security purposes. Recently the Pentagone has used some robots for unmanned aerial vehicles, or UAVs. UAVs help to remove human danger because they are remote controlled planes that are capable of flying themselves if they lose contact with their pilot.

MECHANICS - ROBOT CHASSIS CONSTRUCTION

Robot Part Mounting	
Design	Batteries
Wheels	Electronics
DC Motors	Sensors
Servos	RC Chassis

You should have already read my **build your first robot tutorial** before moving on to this tutorial.

Design

When I first started building my first robot, someone much more experienced than me once said paraphrased, "if you build a mechanically crappy robot with expert programming and control, you will only get a crappy robot; build a mechanically professional robot with crappy programming and control, you will still get a well built robot." Its very good advice which I still use today.

Planning. Would you say someone who plans his future will have a better future? YES! I cannot emphasize any more for you to design your robot out on paper (or computer) first. This means plan out everything, such as what **material to build your robot out of**<, where to put every screw, how you will attach your sensors - EVERYTHING. You will save money and time, and will have a better constructed robot too. To do this, you should draw all your parts out to dimension, mark your holes, and understand how all your parts connect.

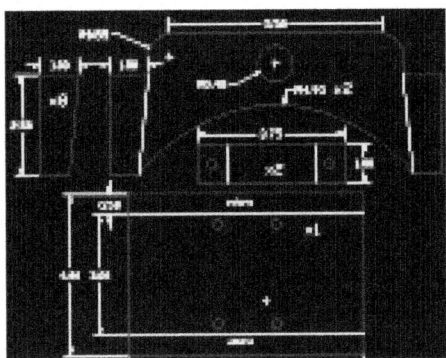

Use fewer and simpler parts! You will probably quickly realize that the fewer parts your robot has, and the simpler they are, the less you will have to design, make, and pay for. As you design your parts, always consider how you will actually manufacture these parts. Don't use unneccessary or over complicated features, or designs you do not have the tools to make or are really hard to make. Fewer and simpler parts also mean a smaller chance for mistakes in your design. K.I.S.S.

Use off-the-shelf parts. When you purchase a part, it costs money. However there is a good chance that the off-the-shelf part is better than anything you can design and build yourself. Off-the-shelf parts are already well engineered, designed, and tested. This means there is less likely to go wrong on your robot, and less time and effort you need to spend to finish it. How much is your time worth? No point in spending 20 hours making a ghetto circuit when you could buy a high quality one for just $20. Ask yourself if the price of a part is greater or less than your willingness and ability to make the part yourself.

Do not use more than 2 or 3 different screw types. There was a robot I once had to work on that another engineer built. He used a different screw type for every part! I had to use like 10 different unlabeled hex wrenches and 2 different screwdrivers to assembly/dissasemble it. Even worse, he had to purchase 10+

different boxes of screws! If you can make your entire robot out of only the very common **4-40** screws, you are on the right track.

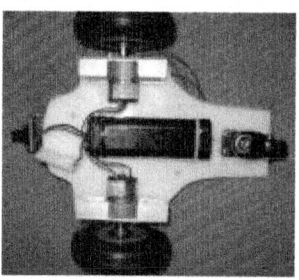

Frame

The frame of your robot is the basic structure to which you attach everything else. It is probably the largest part of your robot, so make sure it is made of a light weight rigid material such as **aluminum** or **HDPE**. I recommend reading those tutorials as well.

Materials

If you are like how I was my first 3 years of building robots, you are probably super cheap and incredibly poor. Robot parts are expensive. Don't even pretend you can make a robot out of parts just lying around your house. Check out my tutorial on **funding your robot**. To summarize, expect to spend about $10-$50 per robot motor, about $20 on frame material, wheels about $8, and about $10 on miscellaneous nuts and bolts. You can try to search garbage bins and other old stuff for this, but you may or not be lucky. You might be able to find a few reasonable good motors, but it might be much harder to find two of the same motors. Just don't build a crappy robot because cereal box cardboard is the only material you can find!

Mounting

When a beginner asks 'how do I attach my robot part?' what they are really asking is 'what is the best way to mount this part?'. Every part on a robot has a different method of mounting. This is due to obvious constraints such as place-ment, weight, size, function, *etc*. So I will go over each of the parts you may want to mount and tell you one or two of the best ways to do this.

Wheel Basics

Wheel diameter. When buying (or making) your wheels you want to put your motor into consideration. For a start, there is **torque** and **velocity**. Large diameter wheels give your robot low torque but high velocity. So if you already have a very strong motor, then you can use wheels with larger diameters. **Servo's** already have good torque, so you should use larger diameter wheels. But if your motor is weak (such as if it does not have any gearing), you want to use a much smaller diameter wheel. This will make your robot slower, but at least it has enough torque to go up a hill! Another dumb mistake someone can make is buying a wheel that has a diameter close to or less than the motor diameter. For example, if you have a 1″ diameter motor, and a 1.5″ diameter wheel, you have a 0.25″ **ground clearance** ((1.5″-1″)/2=.25″). How high is the tallest object you want to go over?

Wheel texture. The texture of your wheel is very terrain dependent. A common mistake for beginners is to ignore the texture of the wheel. If your wheel is **too smooth** then it will not have much friction. This is a serious issue with **omni-wheels**. An all plastic omni-wheel works poorly compared to an omni-wheel that uses rubber for the side wheels. Overly smooth robot wheels would likely skid while accelerating and **braking**. However a wheel that is really rough, such as a foam wheel, has higher friction with the ground leading to innefficiency. You also need to consider **wear and tear** on the wheel.

Wheel width. You do not want it too wide as it causes increased resistance to rotating the wheel on a surface. I once used a 1″ foam wheel on a concrete surface and it was very poor at rotating.

Wheel center hole diameter. This is where you would actually mount the **output motor shaft** to your motor. So you must know the length and diameter of your motor output shaft so that you may put this shaft into the hole of your motor.

Mounting Your Robot Wheel Techniques

Jamming. Lets start with a basic wheel attached to a basic motor. If the wheel does not have a center hole, just drill one out slightly smaller than the diameter of the shaft of your motor. Make sure the hole is perfectly centered!!!! Then you can fill the hole with a little **superglue**, and finally **press fit** (jam, if you will) the motor shaft into the wheel. Perhaps use a little more superglue around the edges.

Servo Wheel Mounting. Mounting a wheel onto a servo is fairly easy, and only requires a little drilling. When you purchase a servo, you also get a bunch of other little goodies with it. One of the items is called a **servo horn**. This is usually a black/red circular, X shaped, or I shaped plastic piece that attaches to the output shaft of the servo. So what you do is attach the wheel to this servo horn, then just screw the servo horn as designed into the servo. In the images below, I drilled two holes into my wheel, 2 holes into the servo horn (the red X thing), and screwed 2 screws into it to hold them together. Then I just attached the servo horn to my servo output shaft with a 3rd screw.

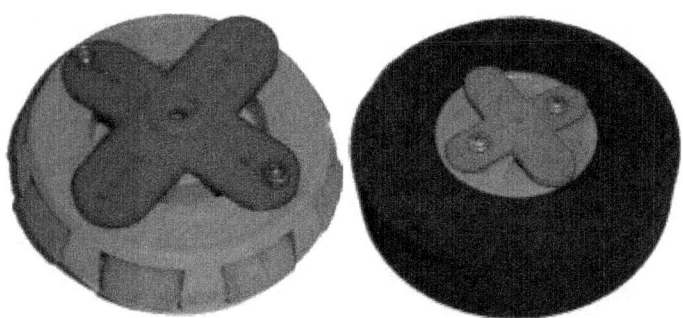

This servo wheel mounting method, with slight modification, can work for other motors too. Just make your own custom 'servo horn' with a custom hole designed to fit your motor output shaft.

Purchasing Wheels

The best robot wheels are actually **hobby RC aircraft wheels**. Strange, huh? Anyway, there is a huge variety of them online for you to choose from. Just go to any RC hobby website. For example, I have bought several wheels from this**online RC hobby store**.

Motors

There are huge varieties in motors. But for your robot I will assume you are using a basic **DC motor**. Make sure you get a motor which already has a **gear box**attached, as it makes your robot much better controlled, more efficient, and stronger. Designing your own gearing and/or chain systems can and will cause you headaches. My first attempt at it was a miserable failure... I recommend against trying it.

Motor Mounting

To mount any type of motor to your chassis you will need to use an **L shaped bracket**. For a DC motor, all you need to do is take a sheet of aluminum, drill two holes in two of the corners, drill two more holes on the other half to match the motor screw holes, then bend the entire piece in a 90 degree angle. This particular image I had found a U shaped piece of aluminum in some bin, I cut it to size, and just drilled the appropriate holes to attach it to my white HDPE chassis.

Servo Mounting

To **mount your servo to your chassis**, once again you will use the little baggie of goodies that came with the servo. You should have two **black cube looking things** with holes in it. There will be three holes in front and a single hole on the bottom. Your servo should also have four mounting holes, two on each side. Using screws, attach the front of the black things to each side of the servo. As shown in the image below, then attach your robot base with two more screws to the bottom hole of the black thingies. This particular robot used a thin reinforced sheet of aluminum as the base with two drilled holes for the mounting screws.

And the completed mounted servo and wheel configuration...

Mounting Robot Batteries

Mounting your **batteries** is very simple. As long as you are using an **RC battery pack** (such as **NiCad** or **NiMH**, you can simply mount your batteries to your robot using strips of **velcro**. This is the velcro that has a sticky tape side and a velcro side. The advantage to this mounting method is that you can easily swap out your batteries for freshly charged ones - great if you are in a **robot competition**. You can also group multiple batteries together using **zip ties**. This is an example of a velcro strip taped to a NiCad battery pack:

Electronics

Mounting your electronics to your robot can potentially be tricky. If you're lucky, your **microcontroller** has holes for screws or something other method of mounting. For this **Cerebellum** microcontroller I drilled holes into my HDPE robot base, then screwed short metal **spacers** into the HDPE. Then all I had to do is put my screws through the Cerebellum screw holes and screw them into the spacers.

Things to keep in mind when mounting your electronics... Keep your electronics high up on your robot. It probably weighs little, so you want the heavier items using the space closer to the ground. You want to keep it out of splashing dirt too. Also keep in mind you probably want certain sensors and other sensitive electronics to stay away from your motors as electronic noise could cause interference. Lastly, plan out your wiring on your chassis - have routing holes drilled through your chassis beforehand, for example.

Mounting Sensors

Mounting sensors is very much a case-by-case basis. What makes mounting sensors difficult is that you are very limited on where you can place them onto your robot. They must be away from noisy motors, and probably at the front/ sides of your robot. You must also keep them protected so that collisions and dirt wont damage your sensors too. If your robot is a line follower, you have to keep your sensors at an exact distance from the ground or risk bad data - even when the ground is uneven! If you are lucky, your sensor has screw holes in just the right place at just the right angle. But in my experience this is very unlikely. Unfortunately I have not yet mastered the art of mounting sensors (because of the huge variety of them), but I can go over a few tricks. So you must get creative, build a mount, use a little glue... you get the picture.

Here you see a homemade aluminum mount for a **sonar**transducer and another mount on a servo for two **Sharp IR Rangefinders**. The sonar mount is aluminum bent at 90 degrees with a few screw holes drilled into it, and plastic spacers to prevent any shorting of the circuit board with the aluminum. The IR

Rangefinder mount is aluminum with **double sided sticky tape** holding the sensors onto the aluminum plate.

Hacking RC Car Frames (The Cheating Method)

Hacking a toy remote controlled car or tank is a great way to get a professionally built robot chassis with much less work. In this image I took a $10 car, hacked the motor wires, and attached all my electronics to it. When buying an RC vehicle, understand that you want a high quality motor so I suggest to avoid the cheap $10 toys. Get something nice if you are going to spend any money on your robot. Notice that you do not need a wireless car! Make sure that the car you will purchase uses a **servo** for it's **rack-and-pinion** steering, or you will have to place in your own. I placed my own servo into this cars' steering mechanism so it's definitely possible, but it took time/effort and wasn't perfect. The servo is the easiest way to do any steering system because it has position control.

Robot Tank Treads

I have never actually built tank treads onto a robot before - mostly because I have been told, and I fully understand, how difficult this task can be. The best way to get treads for your robot is to get a motorized toy tank and hack it into a robot like you would for any RC car. In this robot I ripped off the top part of the tank, drilled screw holes into the plastic, and screwed on a flat sheet of HDPE. From there you can mount whatever you want onto the HDPE - electronics, sensor setups, *etc.* I also ripped out the guts and used the empty space inside to store my happy little NiMH battery. I built this robot in less than 5 hours (minus programming and testing). It's that easy!

APPLICATIONS OF ROBOT MAPPING

The key advantage to autonomous robot mapping is the use of the robot itself. Along with a multitude of sensors, robots can explore dangerous areas that humans would not want to risk their lives exploring. For instance, the military has used robots to map out and patrol hostile areas. If a soldier were to map the same area, he or she could be fired upon by enemy soldiers and killed. The life of that soldier is worth more than even the most expensive robots. Some commercial applications also involve robots mapping dangerous areas.

Robots have been used in nuclear power plants to map out high-radiation areas following nuclear meltdowns. Following September 11, 2001, robots from the University of South Florida were used to explore and map out unstable tunnels in the World Trade Center. Robots have also been used to search volcanoes, meteorites in Antarctica, and map the sea bed. Other possibilities include robots used for mapping and other tasks. Currently there are warehouse systems in design that will map a warehouse, then use the same mapping robots to retrieve inventory that the workers specify.

Autonomous Mapping Methods and Algorithms

Autonomous robot mapping is a complex task that involves a sense of location and a method for exploration. While indoor mapping systems must find a relative frame of reference for location, many outdoor systems utilize GPS to record their absolute location and sensor data. These systems then take this information and integrate it into a spatial map. Because both robotic sensors and GPS have a degree of measurement error, almost all state-of-the-art autonomous mapping systems utilize a Bayesian probability algorithm to infer information about objects and obstacles from the data. Many systems contain algorithms modified for their specific sensor types and sensor performance.

Robot mapping systems must also create an algorithm for exploring areas. These algorithms can range from basic to complex. The most basic algorithm is to start at a point, go straight until an obstacle is reached, and then return to that point and repeat. More complex algorithms may involve using probabilities to estimate where walls or obstacles are. The most efficient way to explore is the use of multiple robots. This allows the mapping to be divided and finished quicker. However, the use of multiple robots adds another layer of design to the system.

Almost all multiple robots mapping systems require the robots to have some sense of relative location to the other robots. The system must also deal with integrating data from multiple robots which can lead to uncertainty when separate robot maps are pieced together. The Distributed Multi-robot Exploration and Mapping Group at University of Washington has designed an innovative system in which multiple robots will map out an area while actively searching for another robot to exchange sensor data with. After this exchange, the robots physically verify their integrated maps for increased accuracy then divide remaining areas for faster exploration.

Typical Mapping Robot Sensors and Parts

A typical robot used for autonomous mapping will require parts for location, movement, and recording data. Most outdoor systems use GPS receivers for locating the robot, while indoor systems use a combination of sensors and reference points. Kiva Systems Inc. has designed an indoor warehouse mapping system that uses barcodes glued to the floor. Indoor robots typically use wheels and motor to move around while a sturdier outdoor mapping robot may need treads such as those found on tanks to navigate bumpy terrain. All mapping robots have a distance sensor such as infrared or sonar and touch. Specialized mapping robots have sensors for their specific application. For example, robots mapping nuclear power plants would have radiation detectors. When multiple robots are used in mapping, a wireless communications medium such as WiFi must be added to each robot.

FIELDS OF APPLICATION

Space Robotics

Fig. : Space Robotics.

The research area Space Robotics deals with the development of intelligent robots for extraterrestrial exploration focusing on:

- Development of robot systems for unstructured, uneven terrain based on biologically inspired innovative locomotion concepts

- Development of multi-functional robot teams usable for different tasks ranging from in-situ examinations to the organisation and maintenance of infrastructure
- Reconfigurable systems for planetary exploration
- AI-based methods for autonomous navigation and mission planning in unknown terrain
- Image evaluation, object recognition and terrain modelling
- AI-based support systems for scientific experiments

Underwater Robotics

Fig. : Underwater Robotics.

This area deals with the development and realization of Artificial Intelligence methods in underwater systems. Main points of research are:

- Development of systems for user support in remote-controlled underwater vehicles employing virtual immersion methods
- Design of methods for autonomous manipulation and mission planning of robot arms in underwater applications, particularly with state-of-the-art sensor technology, such as "Visual Servoing"
- Image evaluation and object recognition with modular and intelligent underwater cameras

- Design of control methods for next-generation autonomous underwater vehicles
- Development of biologically inspired and energy-efficient methods of transport for underwater vehicles, such as oscillating systems

Electric Mobility

Fig. : Electric Mobility: EO smart connecting car.

In the field of electric mobility we are testing concepts for electric vehicles, battery charge technologies, and the collection of vehicle data. We are creating models for intelligent, environmentally sound, and integrated urban mobility. Our research focuses around:

- Development and demonstration of innovative vehicle concepts
- Design of new approaches to mobility and traffic control, application support, technology integration
- Data collection by fleet tests with technologically different electric vehicles
- Coordination of the regional project office of the model region Electric Mobility Bremen/Oldenburg
- Virtualization of the model region, simulation of future, larger vehicle fleets, and predictions of the effects on the model region in terms of traffic volume, infrastructure needs, environmental pollution, and economic efficiency
- Creating a foundation for new business models and traffic concepts on the basis of the data previously collected

Logistics, Production and Consumer (LPC)

Fig. : LPC: Robotlady AILA.

In the area Logistic, Production and Consumer (LPC) new systems are developed which will improve handling and scheduling tasks by using methods of Artificial Intelligence and innovative mechatronic concepts:

- Fast, self-learning image recognition and classification to identify production faults
- "Visual Servoing"-methods to sort piece goods
- Intelligent production memory based on RFID chips in logistic chains
- Handling of deformable piece goods

Search and Rescue (SAR) & Security Robotics

Fig. : SAR: Advanced Security Guard.

In this area, robots will be developed to support rescue and security personnel. Main points of our research are:

- Development of highly mobile platforms for indoor and outdoor applications
- Development of autonomous systems that are able to identify potential victims (SAR) or intruders (Security)
- Development and application of state-of-the-art sensor technology based on radar, laser scanner, and thermal vision to identify objects and persons, resp.
- Embedding of robot systems into existing rescue and security infrastructures
- Autonomous navigation and mission planning

Assistance- and Rehabilitation Systems

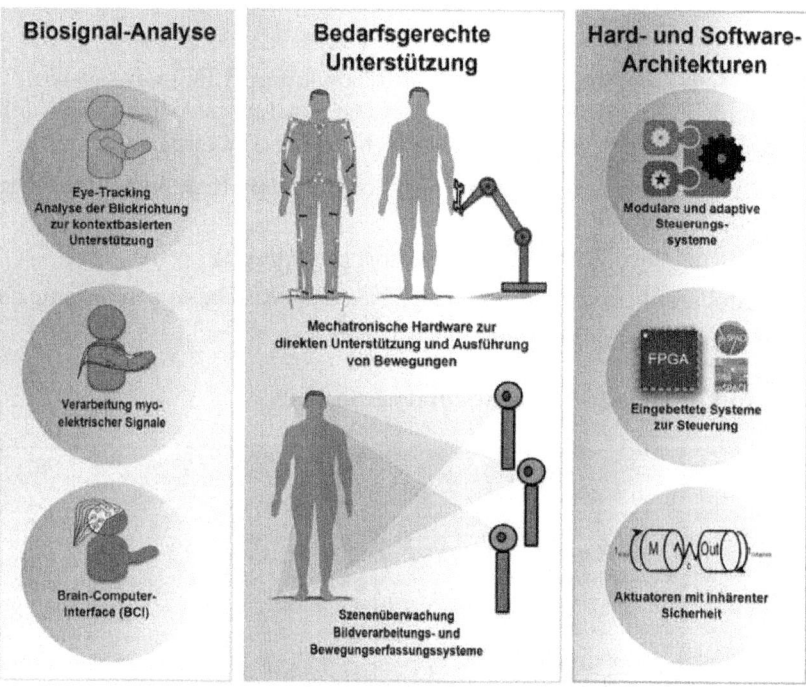

Fig. : Representation of various components within the scope of Assistance- and Rehabilitation Systems

This field deals with robotic systems that can support humans in complex, exhausting or often repeated tasks. Application areas are both help during activites of everyday life (at home or work) and medical rehabilitation. Support can either take place using systems the human is wearing like exoskeletons or orthoses, or by service robots performing the respective task.

Core topics include:

- Concept development, design and construction
- intelligent hardware-system architectures
- software architectures
- embedded biosignal analysis, *e.g.* using information from:
- muscle (EMG)
- eye (eyetracking, EOG)
- or from brain activity (EEG)
- fusion of different sensors
- direct online signal processing (hard- and software)
- robust learning systems capable to adapt
- joint communication layers for better human-machine interaction
- (semi-)autonomously acting systems
- assist-as-needed

Agricultural Robotics

Fig. : Cooperative threshing with several machines.

We develop robots for agricultural applications and transfer methods and algorithms from robotics to conventional agricultural machines. Our objective is to increase the performance of machines and processes and to reduce resource consumption at the same time. Our research is focused on technology applications used in the cultivation of land. Primary research topics are:

- Methods for autonomous planning and navigation of outdoor machinery
- Methods for environmental recognition in agricultural machinery control
- Methods of infield logistics to optimize cooperation and resource consumption between multiple agricultural machines
- Interoperability at the level of communication, processes and knowledge processing

TYPES OF ROBOTS ACCORDING HIS APPLICATION

Various robots are quite simple mechanical machines that perform a dedicated task such as spot welding or assembly operations a repetitive nature task. Besides more complex, multi-task robots systems use sensory systems to gather information needed to control its movement. These sensors provide tactile feedback to the robot so it is able to pick up objects and place them properly, without damaging them. A further robot sensory system might include machine visualization to detect flaws in manufactured supplies. Few robots used to assemble electronic circuit boards can place odd-sized components in the proper location after visually locating positioning marks on the board.

Table : Types of robots according his application.

Industrial Robots are found in a variety of locations including the automobile and manufacturing industries. However, robot technology is relatively new to the industrial scene their roll consists of welding, painting, material handling and assembling.	Robots in Space are name as Remotely Operated Vehicle (ROV). It can be consistent with an unmanned spacecraft that remains in flight or a lander that makes contact with an extraterrestrial body and operates from a stationary position, or a rover that can move over terrain once it has landed.
Educational Robots one example is the Hex Avoider. It is a programmable mobile robot designed to move independently and avoid obstacles. Hex avoider use infrared emitters and receivers to sense its environment. Their roll is demonstrational for teaching basic concepts and gets the attention of future engineers to this field.	Mobile Robots (Transportation) these types of robot operate by control remote deploying sensor position. Their roll consist of sampling payloads, mapping surface and creating a photorealistic 3D models and sent back any kind of visual information of building interiors and any environmental data.
Agricultural Robots one example is the Demeter harvester it contains new controllers, proximity sensors, safeguards and task software specialized to the needs of commercial agriculture processes.	Health Care Robots they are able to perform simple task and improve some medical protocol and procedures.

Degrees of Freedom

A **degree of freedom** is also a term that was cover on page number two and it can be defined as the direction in which a robot moves when a joint is actuated. Each joint usually represent one degree of freedom. Most of the robots used today use five or six degrees of freedom. But this depends on the robot application, for example a pick-and-place application need only three axes specified when a welding robot requires five or six degrees of freedom.

Table : Types of joint links of a manipulator mechanism.

Rotary or revolute joints, these are the most utilized joint and it rotates along the pin as an axis.	Cylindrical joints, these are very rare and are use in some equipment like Parallel Robots or Flying simulator Mechanism.
Spherical joints, these are the third most utilized joint and just slide causing a revolving movement.	Prismatic or Sliding joints, these are the second most employed joint and just slide causing a translation movement.
Screw joints, these just follow the thread of the axis in spiral to move along the axis.	

Robotics Sensors

The word sensor comes from the word sense and it is originate from the *Middle French sens*, sensation, feeling, and mechanism of perception. To improve the performance of the robots it must be able to sense in both ways their internal and external states (the environment) to perform some of the tasks presently done by humans.

As well, much more accurate and intelligent robots are expected to emerge with the newly developed sensors, especially **visual sensors**. Vision provides a robot with a sophisticated sensing mechanism that allows the machine to respond to its environment in an intelligent and flexible manner.

Information Gathered by Robots

First of all, this sensorial perceptions or measurements are gathered by electronic signals, or data that sensors could provide with a limited feedback to the robot so it can do its job. Although proximity, touch, and force sensing play a significant role in the improvement of robot performance. **However, vision is recognized as the most powerful robot sensory capability.**

Robot vision may be defined as the process of extracting, characterizing, and interpreting information from images of a three-dimensional world. This process, also commonly referred to as computer or machine vision, may be subdivided into six principal areas: sensing, preprocessing, segmentation, description, recognition, and interpretation.

Table : Different Type of Sensors.

Proximity sensor senses and indicates the presence of an object within a fixed space near the sensor without physical contact. Different commercially available proximity sensors are suitable for different applications.	**Acoustic sensor** senses and interprets acoustic waves in gas, liquid, or solid. The level of sophistication of sensor interpretation varies among existing acoustic sensors, frequency of acoustic waves and recognition of isolated words in a continuous speech.

Range sensor measures the distance from a reference point to a set of points in the scene. Humans can estimate range values based on visual data by perceptual processes that include comparison of image sizes and projected views of world-object models. Range can be sensed with a pair of TV cameras or sonar transmitters and receivers.	A **force sensor** measures the three components of the force and three components of the torque acting between two objects. In particular, a robot-wrist force sensor measures the components of force and torque between the last link of the robot and its end-effectors by transmitting the deflection of the sensor's compliant sections, which results from the applied force and torque.
A **touch sensor** senses and indicates a physical contact between the object carrying the sensor and another object. The simplest touch sensor is a micro switch. Touch sensors can be used to stop the motion of a robot when its end-effectors make contact with an object.	

ROBOTICS RESEARCH

Much of the research in robotics focuses not on specific industrial tasks, but on investigations into new types of robots, alternative ways to think about or design robots, and new ways to manufacture them but other investigations, such as MIT's cyberflora project, are almost wholly academic.

A first particular new innovation in robot design is the opensourcing of robot-projects. To describe the level of advancement of a robot, the term "Generation Robots" can be used. This term is coined by Professor Hans Moravec, Principal Research Scientist at the Carnegie Mellon University Robotics Institute in describing the near future evolution of robot technology. *First generation* robots, Moravec predicted in 1997, should have an intellectual capacity comparable to perhaps a lizard and should become available by 2010. Because the *first generation* robot would be incapable of learning, however, Moravec predicts that the *second generation* robot would be an improvement over the *first* and become available by 2020, with the intelligence maybe comparable to that of a mouse. The *third generation* robot should have the intelligence comparable to that of a monkey. Though *fourth generation* robots, robots with human intelligence, professor Moravec predicts, would become possible, he does not predict this happening before around 2040 or 2050.

The second is Evolutionary Robots. This is a methodology that uses evolutionary computation to help design robots, especially the body form, or motion and behaviorcontrollers. In a similar way to natural evolution, a large population of robots is allowed to compete in some way, or their ability to perform a task is measured using a fitness function. Those that perform worst are removed from the population, and replaced by a new set, which have new behaviors based on those of the winners. Over time the population improves, and eventually a satisfactory robot may appear. This happens without any direct programming of the robots by the researchers. Researchers use this method both to create better robots, and to explore the nature of evolution. Because the process often requires many generations of robots to be simulated, this technique may be run entirely or mostly in simulation, then tested on real robots once the evolved algorithms are good enough. Currently, there are about 1 million industrial robots toiling around the world, and Japan is the top country having high density of utilizing robots in its manufacturing industry.

Dynamics and Kinematics

The study of motion can be divided into kinematics and dynamics. Direct kinematics refers to the calculation of end effector position, orientation, velocity, and accelerationwhen the corresponding joint values are known. Inverse kinematics refers to the opposite case in which required joint values are calculated for given end effector values, as done in path planning. Some special aspects of kinematics include handling of redundancy (different possibilities of performing the same movement), collision avoidance, andsingularity avoidance.

Once all relevant positions, velocities, and accelerations have been calculated using kinematics, methods from the field of dynamics are used to study the effect of forces upon these movements. Direct dynamics refers to the calculation of accelerations in the robot once the applied forces are known. Direct dynamics is used in computer simulations of the robot. Inverse dynamics refers to the calculation of the actuator forces necessary to create a prescribed end effector acceleration. This information can be used to improve the control algorithms of a robot.

The researchers strive to develop new concepts and strategies, improve existing ones, and improve the interaction between these areas. To do this, criteria for "optimal" performance and ways to optimize design, structure, and control of robots must be developed and implemented.

Education and Training

Robotics engineers design robots, maintain them, develop new applications for them, and conduct research to expand the potential of robotics. Robots have become a popular educational tool in some middle and high schools, as well as in numerous youth summer camps, raising interest in programming, artificial intelligence and robotics among students. First-year computer science courses at several universities now include programming of a robot in addition to traditional software engineering-based coursework. On theTechnion I&M faculty an educational laboratory was established in 1994 by Dr. Jacob Rubinovitz.

Career Training

Universities offer bachelors, masters, and doctoral degrees in the field of robotics. Vocational schools offer robotics training aimed at careers in robotics.

Certification

The Robotics Certification Standards Alliance (RCSA) is an international robotics certification authority that confers various industry- and educational-related robotics certifications.

Summer Robotics Camp

Several national summer camp programs include robotics as part of their core curriculum, including Digital Media Academy, RoboTech, and Cybercamps. In addition, youth summer robotics programs are frequently offered by celebrated museums such as theAmerican Museum of Natural History and The Tech Museum of Innovation in Silicon Valley, CA, just to name a few. An educational robotics lab also exists at the IE & mgmnt Faculty of the Technion. It was created by Dr. Jacob Rubinovitz.

Robotics Afterschool Programs

Many schools across the country are beginning to add robotics programs to their after school curriculum. Two main programs for afterschool robotics are Botball and FIRST Robotics Competition.

Employment

Robotics is an essential component in many modern manufacturing environments. As factories increase their use of robots, the number of robotics–related jobs grow and have been observed to be steadily rising.

NEW APPLICATIONS FOR MOBILE ROBOTS

Mobility promises to be the next frontier in flexible robotics. While fixed robots will always have a place in manufacturing, augmenting traditional robots with mobile robots promises additional flexibility to end-users in new applications. These applications include medical and surgical uses, personal assistance, security, warehouse and distribution applications, as well as ocean and space exploration.

"We see increased interest in mobile robotics across all industries. The ability of one mobile robot to service several locations and perform a greatly expanded range of tasks offers a great appeal for specialized applications," says Corey Ryan, Medical Account Manager at KUKA Robotics Corp.

Autonomous mobile robot on the
job, courtesy Adept Technology Inc.

Mobile Apps

Mobile robots are proliferating says Rush LaSelle, Vice President and General Manager with Adept Technology Inc. (Pleasanton, California). "In the industrial space, mobile robots are redefining the playing field for autonomous guided vehicles (AGVs) in that modern mobile platforms are capable of operating in areas without requiring alterations or investment into existing infrastructure. Mobile robots overcome a historical impediment of AGVs, their inability to dynamically reroute themselves. Mobile robots are outfitted with advanced sensory and enhanced intelligence systems."

Reduced costs enable deploying both large and small fleets of vehicles in warehouse distribution and line-side logistics applications, LaSelle adds.

Mobile robots can be particularly useful in painting and de-painting applications, says Erik Nieves, Director of Technology in the Motoman Robotics Division of Yaskawa America Inc. (Miamisburg, Ohio). "Mobility is a force multiplier for robots and I see that in de-painting very large structures such as C-130 aircraft. Two fixed robots cannot de-paint an entire aircraft between them because they cannot reach everywhere." More than two fixed robots constitutes too much hardware with very little throughput. "Each robot is painting a little piece then sit idle, parked more than moving," says Nieves.

Nieves suggests that rather than adding additional fixed robots around the aircraft, end-users needs a way to have two robots deal with an entire aircraft. "To de-paint an entire aircraft with two robots, those two robots need to move." Putting the robots on servo tracks or a gantry is unfeasible due to aircraft's geometry. "Putting two seven-axis robots on mobile platforms and driving them around the aircraft" is a better solution, Nieves says.

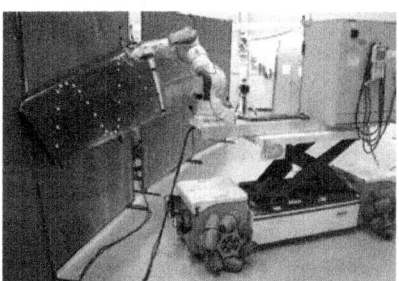

Mobile robot working on aircraft wing, courtesy
Southwest Research Institute.

Likewise, Paul Hvass, Senior Research Engineer with the Southwest Research Institute (SwRI, San Antonio, Texas) says mobile robots facilitate cost-effective paint removal from large aircraft. "The motivation behind the development of our Metrology-Referenced Roving Accurate Manipulator (MR ROAM) was to demonstrate high-accuracy, industrial-grade mobile manipulation for very large workspaces, an enabling capability for applications like aircraft paint stripping. SwRI has a 25-year history of developing, deploying, and supporting custom robots for fighter jet paint stripping and other large scale applications." Hvass goes on to say, "To economically strip paint from larger planes, mobile automation is needed. In the future, we envision mobile robots developed for large-scale tasks including aerospace, off-shore, and road, bridge, and building construction. These robots will initially undertake light-duty tasks such as painting, cleaning, and inspection before moving on to heavier-duty tasks as mobile robotic technology matures," Hvass concludes.

Medical/Surgical Applications

Corey Ryan talks about potential uses of mobile robotics in medical and other life sciences applications. "Medical applications are always a growing field with huge untapped applications like drug delivery, or the development of mobile treatment systems for specialized equipment."

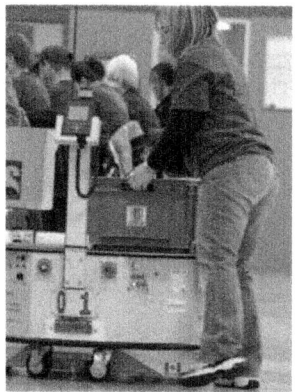

People and mobile robots working collaboratively, courtesy RMT Robotics Ltd.

Autonomous mobile robots (AMR) can play a role in assisting doctors in surgical procedures, says, Bill Torrens, Director of Sales and Marketing with RMT Robotics Ltd. (Grimsby, Ontario, Canada). "AMR technology is applied in surgical applications. Based on inputs, the robot arm assists the surgeon to perform a task. Path-planning algorithms move the robot autonomously."

Sean Thompson, Applications Engineer at MICROMO (Clearwater, Florida) sees an increase use of robotics for automated prosthesis fabrication. "Minimizing motor size helps make prostheses more related to the natural human form. That comes down to applying power to build prostheses that more closely emulates the body's natural capabilities."

Danger Seeker

Mobile robots can access areas dangerous to humans, says, Andrew Goldenberg, President of Engineering Services Inc. (ESI, Toronto, Ontario, Canada). "Mobile robots are used to reach inaccessible areas such as nuclear power plants. Mobile robotics are very useful in nuclear environments with high levels of radiation, particularly during a disaster or threat of a disaster."

Goldenberg goes on to say, "Some companies are using robotics underwater while others want to develop robotics for military applications, shoreline exploration of mines, and for repairing a ship's structure." ESI is involved with mobile robots for space exploration, such as rovers remotely moving on Mars.

Mobile robot bristling with sensors on tracks,
courtesy Engineering Services Inc.

As a caveat, Goldenberg says, "Current robotics are not quite sufficiently designed to withstand high radiation affecting their electronic circuitry. Some attempts to design mobile robotics specifically for use in this environment have been made."

Wireless communication with mobile robots is still a challenge, says Goldenberg. "If mobile robots go underground or in areas of low connectivity like subway tunnels, control of the robot could be lost."

Hvass also talks about communication to and from mobile robots. "If the robot communicates with infrastructure over a wireless link, that link is vulnerable due to bandwidth sharing, variable distances between radios, obstructions, and non-deterministic protocols."

Mobile robots for use in inaccessible areas is also on the mind of Sean Thompson. "We see more interest in undersea robotics with smaller non-tethered robots used by research facilities. Aerial robotics tends to go either way, smaller platforms and larger platforms, depending on the mission. Camera packages have gotten smaller which allow aerial robots to roam at lower altitudes in shorter distances on smaller aircraft. These remote-controlled aircraft are collecting highly-detailed and accurate video."

Thompson speaks of other military applications of mobile robotics. "Troopers could carry heavier loads with robotic pack dogs and exoskeletons. This technology is different from replacing a service dog but will be commonplace in five to 10 years."

LaSelle also sees mobile robotics utilized for patrol and monitoring applications. "Another key expansion of mobile robotics has been in monitoring, security and patrolling. Patrolling applications provide users with the ability to monitor intrusion, thermal and other environmental conditions. A key area of activity has been the monitoring and patrol of vacant properties as well as warehousing spaces." This increased ability is due to the reliability and low costs attributed to autonomous vehicle patrol capabilities, LaSelle says.

Thermal monitoring is of special interest to Internet server farms and other sensitive electronic or mechatronic systems. Water ingress is also commonly monitored by way of mobile robotics, LaSelle notes.

Mobile robots are finding their way into other non-industrial applications. "The reduced cost of deployment and ownership mobile robots have extended their reach into non-factory applications. The current generation of smart vehicles is leading hospitals, laboratories, and some offices to employ mobile robots to alleviate the use of skilled labor for mundane transport tasks."

Continuing, LaSelle adds, "Mobility is already the norm in service applications and this sector is primed for tremendous growth. Service robotics is expected to overshadow the industrial robot sector in a matter of a few years. Adept believes mobile robots will be an exciting area in coming years," reports LaSelle.

Mobility=Lean

The vision of truly lean manufacturing is being realized through mobile robotics says Torrens. "Mobile robotics connect islands of automation. The last frontier of lean manufacturing facilitates the connection between manufacturing work cells. Mobile robots are now used for transporting materials from donation areas and taking these raw materials to a work cell."

Torrens says mobile robotics provides a much higher level of flexibility for manufacturers. "For example, a manufacturing facility normally delivers a bin of 100 parts for a machine to work on. This is an example of batch processing, not lean manufacturing. Lean manufacturing embraces a piece-work philosophy, or a smaller batch philosophy. If taking one piece at a time to a machine, manufacturers have more flexibility with robotic transport between manufacturing cells. That approach is lean manufacturing as originally intended."

Torrens believes "mobile robots have finally achieved the goals of what the factory of the future was supposed to look like. The machines were in place but the transport logistic was not." Mobile robotics provides that logistical support, argues Torrens. "To realize lean manufacturing, robots must be highly intelligent and able to autonomously deliver parts from any random origin to any random destination. Mobile robot technology up to this point has been unable to deliver materials in a just-in-time way."

LaSelle anticipates mobile robotics serving the ends of lean manufacturing through processing of optimal batch sizes in warehouse and palletizing applications. "Adept sees the combination of mobility and manipulation as a powerful combination as evident in the increasing demand for case-picking applications. Companies want to move smaller batch sizes throughout their facilities." End-users want to move less than a full pallet from a warehouse to a production line, concludes LaSelle.

"Companies look for solutions to pick cases or parts individually within a warehouse as compared to pulling a full rack. As this trend continues, expect to see more demand for systems encompassing mobility, manipulation, and vision. Given the rate of technological advancement and drive for smaller batch sizes in manufacturing, we will see mobile robots become a staple in a large cross section of manufacturing within the next six to seven years," foresees LaSelle.

Autonomous Locomotion

Genuine independent mobility is necessary for robotics to add significant value to manufacturing says Erik Nieves. "Mobility moves robots from being machines to production partners. The robot has to move to the work but if the robot is bolted to the floor and has no work before that robot, the robot is adding zero value to the production process." Bringing a mobile robot to where production is rather than bringing production to a fixed robot is the philosophical underpinning of mobile robotics, Nieves says.

Any mobile platform must address issues relating to power, navigation, and calibration, says Nieves. "Instead of mobile robots tethered to a source of power through an umbilical, the robot will dock to a power source when reaching a point of interest, to recharge while working." On-board power simply keeps the robot mobile during transit.

Nieves turns his attention to navigation, or "How the robot gets from A to B autonomously. Using simultaneous localization and mapping, the mobile robot can go from one station to the next largely on its own with without many changes to the facility. To change the mobile robot's path, [a number of guidance] labels are put somewhere else," describes Nieves.

Calibration, the final element in Nieves' approach, is a measure of how close the robot gets to it intended destination. "The robot must calibrate itself to the machine in front of it every time it arrives at one. Calibration is done by some means, such as touching off on three points or using a vision sensor to allow the robot to determine its location."

BIOLOGICALLY INSPIRED MACHINES: MAPPING AND LOCALIZATION

The task of building a robot capable of concurrently mapping and localizing (CML) in an unexplored and dynamic environment is an outstanding problem in the field of robotic autonomy. Although there are implementations of CML that have solved the problem, most have relied on small environments or time and memory constraints that are not be representative of the real world. For example, robots using an Extended Kalman Filter (EKF) are able to correctly perform CML, but are unable to handle the Relocation Problem, when the robot is kidnapped and moved and must then continue running CML on the original map. Different implementations that seek to handle Relocation often cannot work in real-time, or rely on an extensive memory, limiting the size of the map that they can handle.

While most research to the problems in CML lies in further enhancements of the existing algorithms, this chapter looks for inspiration in the natural world. Human mechanisms for handling the relocation problem and limited memory capacity are real-time functions. These models of human memory can be drawn analogously into an existing implementation of CML, and provide the robustness that is currently lacking.

This chapter describes an existing implementation of Simultaneous Localization and Mapping (SLAM) that is capable, but fundamentally limited. The existing implementation can only handle a small number of features in the environment and is unable to handle the relocation problem. The two biologically inspired enhancements to the SLAM implementation provide the following new capabilities:

- Handle the relocation problem. If the robot is abruptly moved to a new place in the map, it undergoes a real-time, sample-based search of its map to determine its new position and orientation before continuing its plan. This mirrors the biological recognition system that allows humans to recognize and readjust to places they have previously seen.

- Handle larger environments with memory management. As the robot explores, certain stored features in the map will slowly decay until they reach a threshold, when they are forgotten. As a result, the robot keeps sparser maps of areas it has not seen in a while, and more detailed maps of more recent areas. This allows the robot to maintain old information, while still adding to the map. This mirrors the biological memory storage system where humans remember old facts, but not with the same depth and richness as more current information.

Using these two biological systems, a simulated running of the enhanced SLAM implementation is able to handle being kidnapped as well as environments with many times more features.

The original SLAM implementation, both biological improvements and experimentation that shows the added robustness of the new features. The SLAM implementation, including how observations were made and stored, as well as how it solved the basic CML problem. The descriptions of the Recognition and Memory Storage enhancements described briefly above. Our experimental results after running the robot on a simulated environment with and without the biological enhancements.

SLAM Implementation

The initial SLAM implementation used a system of linearized stochastic mapping to handle the problem of CML. The robot was given a movement plan for exploration and at each step along the way, took observations of the environment with a limited range scan. Observations were matched to features and the robot updated its map and localized itself using a Kalman Filter.

The actual task of observation and preprocessing the feature information is dependent on an artificially simplified simulation environment. Each feature is given a unique identifier. When the robot scans for observations, it measures the distances to a set of features within its scan range, with some arbitrary error adjustment to simulate real-world noise. However, it also receives the unique identifier of each feature, which is unrealistic in a real world setting, but adequate for the simulated intents of this SLAM implementation. The robot compare the feature ID against its stored map, and determines whether or not it is a new feature. If it is in fact a new feature, it is added to the map. If it is not, the robot has to update the map.

The heart of stochastic mapping lies in the second stage where the features are actually processed. This implementation relies on a Kalman Filter Update to handle the necessary computation of updating the map and localizing the robot. The KF approach relies on creating a Gaussian distribution of the predicted state of the environment. Using the measurements from a feature observation, a Bayes filter is used to correct the prediction and update the state.

Using a simple loop of motion, prediction, observation, update, the SLAM algorithm is able to successfully map out a feature-laden environment. Given the random observation errors, the first-pass map is within 2 standard deviations of the correct map.

Recognition

A memory recognition system that will allow the existing SLAM implementation to handle the relocation problem, when the robot is moved in an unknown region of the map. Since the existing SLAM implementation relied on feature identification, which makes the relocation problem trivial, we ignore feature ID's when observing and processing features in memory recognition. Instead we give additional quantitative value to the feature, an angle of orientation and length that is observable with a degree of error comparable to those in the distance errors. This additional feature information is necessary for narrowing down the choice of possible observable features during recognition.

The recognition technique relies on searching the geometric constraints of the observable features to determine the most likely position and orientation of the robot. In a standard stochastic mapping this search involves matching each observation to a possible feature, and searching through all combinations of observation-feature pairs for the optimal fit. This technique is time inefficient as searching through the feature space is exponential in the number of measurements. This chapter draws on a new technique to minimize the search time by limiting the size of the search tree.

Geometric Constraints

In order to reduce the size of the observation-feature search tree, we use three different types of geometric constraints to determine the likelihood of an observation-feature pairing. The first of these constraints is a *unary* test of the feature's angle of orientation with respect to the observed measurement of that angle. We determine the stochastic parameter vectors and a correlation matrix for the feature and the observed angle. Then, using a chi square test for statistical significance we can decide if the chosen pair should be considered or ruled out.

The second geometric constraint we consider is a *binary* test comparing the relative geometry's between two feature-observation pairs. In this case we create a vector of the difference between the distances of the features and the observed measurements. Using the corresponding correlation matrix as before, we arrive at a bound of probability for the chi square test.

The final constraint is the *joint compatibility* test. This test checks the geometric relationship among all the feature-observation pairs. The intent behind this test is to rule out the cases where all subsets of two pairs pass the *binary* test, but the set, as a whole is geometrically infeasible. The *joint computability* test computes the chi square test on all the combined pairs. This test makes sure that the inclusion of spurious observations are minimized, since hypotheses with spurious observations will not pass the statistical test.

Branch & Bound

Using the three geometric constraints, we implemented a recursive *branch and bound* search algorithm that returns the best set of feature-observation pairs. The joint computability branch and bound (JCBB) algorithm works as follows:

1. If all observations have been mapped:
    *   Count the number of feature-observation pairs in the current set, H, and in the previous best set, B.
    *   Designate the set with more pairs as the new B.
2. If the current observation, o, has not yet been mapped to a feature:
    *   For every feature, fi, in the map:
    *   If unary(o, fi) AND joint_comp([H, fi]:
    *   JCBB([H fi])
    *   If no feature could be mapped to o AND there are more pairs in H than in B:
    *   JCBB({H 0]) //(spurious observation)

Running this algorithm over the set of observations made during attempted recognition is considerably more efficient than the previous attempts to search the entire tree of possible sets of observation-feature pairs. The JCBB algorithm relies on eliminating entire branches as soon as a pair is added which violates the global optimality of the pair set.

Random Sampling

Although the JCBB algorithm is a dramatic improvement the full tree search, it still requires a lot of computationally expensive geometric constraint functions.

In the case of a move to an unexplored section of the map, the entire search tree still must be explored. One simple heuristic to improve both the worst case and general case search time is to split the search into two parts – a guess and a proof, using a random subset of observations for each part.

When the robot is relocated, it makes n observations of its surroundings. Using the *random sample* improvement to the JCBB, we select p of these measurements to generate a hypothesis of the new location using only their *unary* and *binary* constraints. Without using the expensive *joint compatibility* test, we reduce the search tree to a significantly smaller hypothesis. Then, we run JCBB on the remaining observations to verify the accuracy of the hypothesis.

Memory Limitations

The biologically inspired memory functions, which allow the SLAM implementation to act more robustly in large environments. The SLAM code as originally implemented was only intended to handle a relatively small number of features. This limit is reflective of two problems in CML in actual robots:

(1) limited physical memory, and

(2) matching observations to features becomes exponentially harder and more computationally costly as more features are stored.

In order to solve these problems, we implemented a memory management system to handle the resources more efficiently.

When a feature is added to our memory, it is assigned a *freshness* function with probability p. The freshness is equivalent to a time stamp of feature acquisition. Over time, the freshness will decay and eventually drop below a threshold, at which point it is available to be removed from memory. Similarly, if the feature is observed again, its freshness will return to the initial Peak State. Features will be removed only when an upper limit of features to be held is reached. Furthermore, a feature that is revisited many times will be updated such that it will reflect its robustness in the explored environment.

The result of this new dynamic memory is manifold. Since only a random fraction of features are given freshness functions, the map loses some of its detail, but will retain its general shape. Furthermore, the areas that the robot has most recently explored will be the "freshest" and most densely mapped. Regions that have been mapped in the past, and may not be seen again for a long time, are still kept in memory, much more sparsely.

Experimentation

To test the relocation and dynamic memory management system we have used an artificial dataset of an environment similar to a the inside of a building and a vehicle that has an odometry sensor mounted on it. In the beginning of a simulation the vehicle is given a constant velocity and directions as a function of time. The vehicle performs CML while traveling in its environment. The depicts concurrent mapping and localization in the simulated environment. The vehicle is able to reliably map the features in its environment and travel through it, compensating for noise contributed by measurements and vehicle controls.

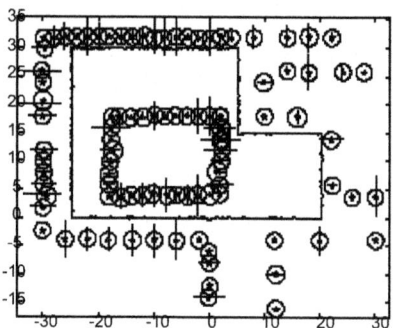

Fig. : CML map obtained by the vehicle. The continuous and dashed lines depict the vehicle actual and planned trajectories, respectively. A star marks each feature location, and the associated bar is a unique characteristic of the feature (determined by angle and length). Finally, the error in estimation in vehicle trajectory is marked by the deviation of the continuous line from the dashed line, and the error in estimate the feature location is marked by the radius of the circle.

First we tested the relocation implementation. We ran CML continuously until the environment was mapped and then kidnapped the robot. The algorithm was able to map the environment based on very few sensor readings at each step (as few as 3 observations). The ability to map the environment is based mainly on the relationship between the measurements, vehicle position and noise in the system. However, for the recognition to work, the features should demonstrate both intrinsic uniqueness (*e.g.*, whether a feature is a corner or not) and correlations between features that is independent of the vehicle position.

That is, the vehicle should be able to uniquely determine its location from a set of features available in its immediate location. Table demonstrates the dramatic effect of distinctiveness on recognition performance. Changing the length and angle of the bar to correspond to walls and corners simulated distinctiveness of features. For example, all the bars with length close to zero while the other features correspond to walls. Distinctiveness was manipulated by add noise to these characteristics (not shown here). Importantly, the features distinctiveness plays a major role in the successful estimation of the new location.

	Relocation Estimation Error	
Features	X	y
Normal	-1.30 ± 20.73	-0.58 ± 26.74
Distinct	-0.12 ±.41	0.09 ± 0.39

Table: Relocation estimation error (±SD) as a function of feature distinctiveness. Note that the distinctive features show a much lower error and variance.

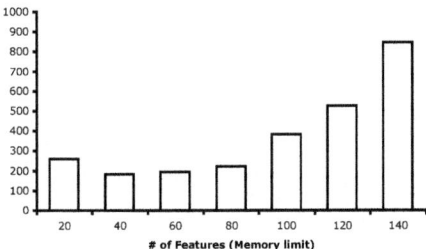

Fig. : Mean execution time vs. memory size.

Secondly, we tested the CML with memory limitations. In order to test the effects of using dynamic memory management on the performance of the mapping we measured the error in mapping as a function of number of features and the time needed to perform the exploration. Although the algorithm is successful in mapping the environment with a limit on the number of features and it does so faster then it would do without the memory limitation, it does have larger variance and errors in the resulting map when the memory is limited. Nevertheless, the features' variance does converge in this case and therefore the features do not contribute noise into the mapping and the CML is successful.

Finally, we tested whether the relocation is still effective when using dynamic memory management. The relocation is successful even when memory limitations are applied.

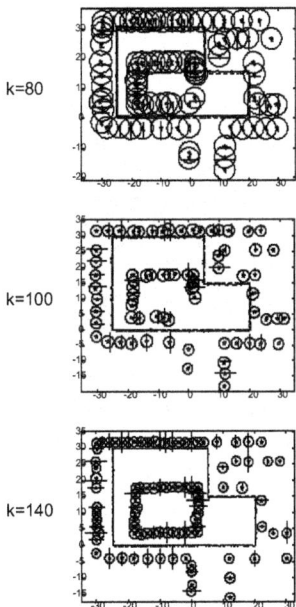

Fig. : Feature estimation as a function of memory limit. Three maps with different maximum number of features are shown from lowest to highest (top to bottom). Note that the radius of the circles, and thus the error in estimating their location is inversely related to the memory size.

Chapter 4

MOBILE ROBOTICS TECHNOLOGY

Robotic arms are relatively easy to build and program because they only operate within a confined area. Things get a bit trickier when you send a robot out into the world.

The first obstacle is to give the robot a working locomotion system. If the robot will only need to move over smooth ground, wheels or tracks are the best option. Wheels and tracks can also work on rougher terrain if they are big enough. But robot designers often look to legs instead, because they are more adaptable. Building legged robots also helps researchers understand natural locomotion — it's a useful exercise in biological research.

Fig. : Fujitsu's HOAP-1 robot.

Typically, hydraulic or pneumatic pistons move robot legs back and forth. The pistons attach to different leg segments just likemuscles attach to different bones. It's a real trick getting all these pistons to work together properly. As a baby, your brain had to figure out exactly the right combination of muscle contractions to walk upright without falling over. Similarly, a robot designer has to figure out the right combination of piston movements involved in walking and program this information into the robot's computer. Many mobile robots have a built-in balance system (a collection of gyroscopes, for example) that tells the computer when it needs to correct its movements.

Fig. : NASA's Frogbot uses springs, linkages and motors to hop from place to place.

Bipedal locomotion (walking on two legs) is inherently unstable, which makes it very difficult to implement in robots. To create more stable robot walkers, designers commonly look to the animal world, specifically insects. Six-legged insects have exceptionally good balance, and they adapt well to a wide variety of terrain.

Some mobile robots are controlled by remote — a human tells them what to do and when to do it. The remote control might communicate with the robot through an attached wire, or using radio or infrared signals. Remote robots, often called puppet robots, are useful for exploring dangerous or inaccessible environments, such as the deep sea or inside a volcano. Some robots are only partially controlled by remote. For example, the operator might direct the robot to go to a certain spot, but not steer it there — the robot would find its own way.

MOBILE ROBOT

A mobile robot is an automatic machine that is capable of movement in any given environment.

Fig. : A spying robot is an example of a mobile robot capable of movement in a given environment.

Mobile robots have the capability to move around in their environment and are not fixed to one physical location. In contrast, industrial robots usually consist of a jointed arm (multi-linked manipulator) and gripper assembly (or end effector) that is attached to a fixed surface.

Mobile robots are a major focus of current research and almost every major university has one or more labs that focus on mobile robot research. Mobile robots are also found in industry, military and security environments. Domestic robots are consumer products, including entertainment robots and those that perform certain household tasks such as vacuuming or gardening.

Classification

Mobile robots may be classified by:

- The environment in which they travel:
    1. Land or home robots are usually referred to as Unmanned Ground Vehicles (UGVs). They are most commonly wheeled or tracked, but also include legged robots with two or more legs (humanoid, or resembling animals or insects).
    2. Aerial robots are usually referred to as Unmanned Aerial Vehicles (UAVs)
    3. Underwater robots are usually called autonomous underwater vehicles (AUVs)
    4. Polar robots, designed to navigate icy, crevasse filled environments
- The device they use to move, mainly:
    1. Legged robot : human-like legs (*i.e.* an android) or animal-like legs.
    2. Wheeled robot.
    3. Tracks.

Mobile Robot Navigation

There are many types of mobile robot navigation:

Manual Remote or Tele-op

A manually teleoperated robot is totally under control of a driver with a joystick or other control device. The device may be plugged directly into the robot, may be a wireless joystick, or may be an accessory to a wireless computer or other controller. A tele-op'd robot is typically used to keep the operator out of harm's way. Examples of manual remote robots include Robotics Design's ANATROLLER ARI-100 and ARI-50, Foster-Miller's Talon, iRobot's PackBot, and KumoTek's MK-705 Roosterbot.

Guarded Tele-op

A guarded tele-op robot has the ability to sense and avoid obstacles but will otherwise navigate as driven, like a robot under manual tele-op. Few if any mobile robots offer only guarded tele-op.

Line-following Robot

Some of the earliest Automated Guided Vehicles (AGVs) were line following mobile robots. They might follow a visual line painted or embedded in the floor or ceiling or an electrical wire in the floor. Most of these robots operated a simple "keep the line in the center sensor" algorithm. They could not circumnavigate obstacles; they just stopped and waited when something blocked their path. Many examples of such vehicles are still sold, by Transbotics, FMC, Egemin, HK Systems and many other companies.

Autonomously Randomized Robot

Autonomous robots with random motion basically bounce off walls, whether those walls are sensed

Autonomously Guided Robot

Fig. : Robot developers use ready-made autonomous bases and software to design robot applications quickly. Shells shaped like people or cartoon characters may cover the base to disguise it. *Courtesy of MobileRobots Inc.*

An autonomously guided robot knows at least some information about where it is and how to reach various goals and or waypoints along the way. "Localization" or knowledge of its current location, is calculated by one or more means, using sensors such motor encoders, vision, Stereopsis, lasers and global positioning systems. Positioning systems often use triangulation, relative position and/ or Monte-Carlo/Markov localization to determine the location and orientation of the platform, from which it can plan a path to its next waypoint or goal. It can gather sensor readings that are time- and location-stamped, so that a hospital, for instance, can know exactly when and where radiation levels exceeded permissible levels. Such robots are often part of the wireless enterprise network, interfaced with other sensing and control systems in the building.

For instance, the PatrolBot security robot responds to alarms, operates elevators and notifies the command center when an incident arises. Other autonomously guided robots include the SpeciMinder and the Tug delivery robots for hospital labs, though the latter actually has people at the ready to drive the robot remotely when its autonomy fails. The Tug sends a letter to its tech support person, who then takes the helm and steers it over the Internet by looking through a camera low in the base of the robot.

Sliding Autonomy

More capable robots combine multiple levels of navigation under a system called sliding autonomy. Most autonomously guided robots, such as the HelpMate hospital robot, also offer a manual mode. The Motivity autonomous robot operating system, which is used in the ADAM, PatrolBot, SpeciMinder, MapperBot and a number of other robots, offers full sliding autonomy, from manual to guarded to autonomous modes.

MOBILE WHEELED ROBOT FOR MECHATRONIC CONTEST

The mobile robot designed for a line following competition and obstacles avoidance. For this type of mobile robot tasks looks for energetic and informational autonomy. Informational autonomy refers to the ability of the robot to perform the task for which it was programmed and can make decisions in extreme situations. Line following contests usually consists in a sinuous black path, 15mm or 19mm wide on a white background with arcs with a minimum radius of 100mm and intersections at right angles. In some cases, there may route markings adjacent track line perpendicular to it, which are placed at the beginning of each arc. These marks are used for robots to adapt the speed while cornering.

Robot Task Definition

The robot, which size should not exceed 200mm x 200mm x 200mm, must be able to travel independently in short time, two types of routes: a speed one and an obstacle one. Speed route is a route with simple, straight lines and arcs of minimum radius of 100mm, without junctions or line breaks. Obstacle courses consists of straight lines, arcs of minimum radius of 100mm and the obstacles that intersection at right angle turns at right angles, consecutive series of turns at right angles to form a step ladder with 30mm, roundabouts, line stopping and resuming it after 50mm in any direction and line break and replace it with an auxiliary wall.

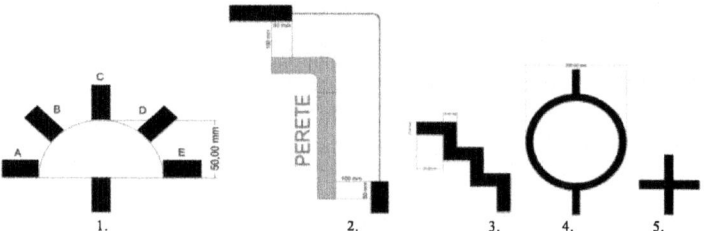

Fig. : Obstacle types.

Line Follower Robot Mechatronic Structure

The proposed robot is a wheeled mobile robot with two differential type wheels and a rear one for support because it is the most effective type of robot for this task. The chassis is constructed of steel sheet to ensure the necessary rigidity at low cost. On this chassis are fixed motors and wheels, support wheel, electronic circuits, battery and front panel sensor. All components are mounted as to provide a center of gravity located low down and close to the axis wheels. 1331T012SR motors used are from MINIMOTOR, coupled with planetary gear transmission ratio of 14:1 and incremental rotary transducers.

Fig. : Mobile robot for line following competitions: 3D – CAD (left) and the final product (right).

In terms of control electronics, microcontroller Atmel ATmega 2560 used is mounted on a development board Arduino Mega. The development board was chosen because it contains all the necessary electronics for the operation and programming of microcontroller and Arduino programming environment contains many powerful libraries that simplify complex programming tasks.

For motor drives was used a L298 chip from STMicroelectronics which serves dual H-bridge. The battery used is composed of 3 cell lithium polymer capacity 2Ah. It provides a nominal voltage of 11.1 V. For line detection, the robot is equipped with reflective sensors. The sensors used are manufactured by Parallax QTI type.

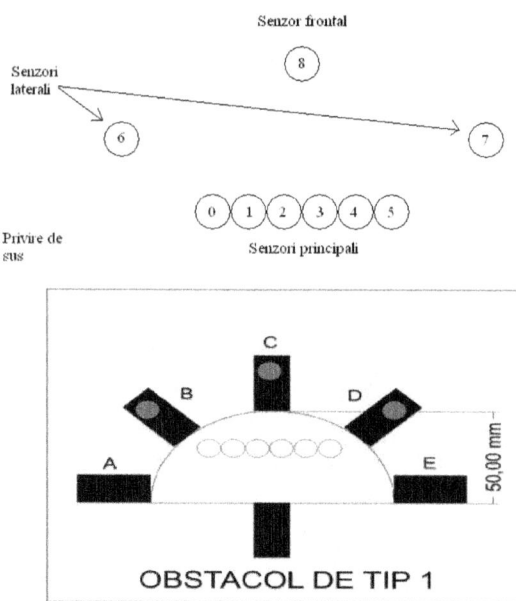

Fig. : Sensors placement and an example for one kind of an obstacle.

The controllers are implemented on the microcontroller. It receives analog signals from the line detection sensors and, using analog-digital convertor (ADC), the values are converted into digital values which then will be processed. The motor rotary encoder produces two digital signals that are also processed by the microcontroller. The results of processing are two type pulse width signals and two digital signals for reversal. These four signals are fed into an integrated circuit L298 dual H-bridge type which supplies the motors. The general block diagram has another block regarding wall obstacle strategy, in fact reading of the specialised sensor.

Mathematical Model

Starting from the position of gravity centre and the reaction between wheels and ground terrain, some calculations were made in order to know the values for actuation force at the ground level function of the nominal battery voltage Ubat and duty cycle fu. Those values are very important for the command strategy.

$$U_{med} = f_u \cdot U_{bat}$$

$$I = \frac{U_{med}}{R}$$

$$M_m = \left((k_m \cdot i) - M_R\right) \cdot \frac{1}{red} \cdot \eta_{red} \cdot \eta_t$$

$$M_a = M_m - M_f$$

$$F_a = \frac{M_a}{r} \cdot 2$$

Where: U_{med} - average voltage, R- rotor resistance, M_m - the motor torque at the axis of the actuating wheel, i – motor current, M_R – friction torque at the motor bearings level, red – gear transmission coefficient, η_t - 1:1 transmission efficiency, η_{red} - gear efficiency, M_a – actuating torque for actuating wheels, M_f is friction torque, r is the wheel radius, F_a – actuating force for actuating wheels.

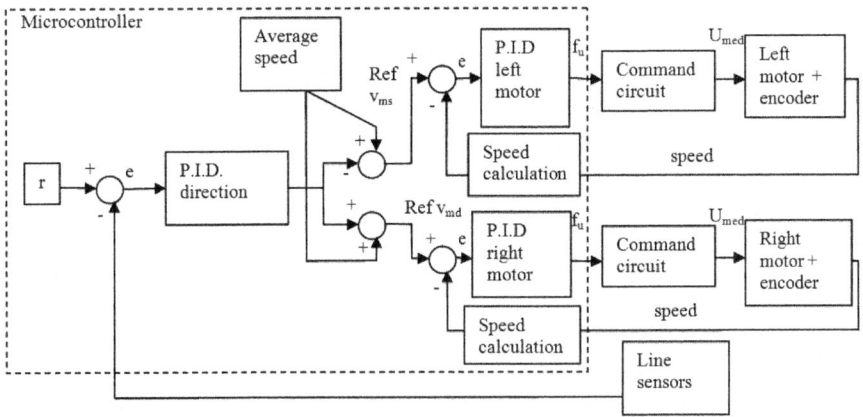

Fig. : Control block diagram for the speed contest.

Where: Umed - average voltage, R- rotor resistance, Mm - the motor torque at the axis of the actuating wheel, i – motor current, MR – friction torque at the motor bearings level, red – gear transmission coefficient, - 1:1 transmission efficiency, - gear efficiency, Ma – actuating torque for actuating wheels, Mf is friction torque, r is the wheel radius, Fa – actuating force for actuating wheels.

By developing a mathematical model of the robot implemented in Matlab / Simulink environment, the device behavior was observed for various speeds, analyzing the power consumption and operating force developed. In this way were achieved an optimization of control.

DOMESTIC ROBOT

A domestic robot, or service robot, is an autonomous robot that is used for household chores. Thus far, there are only a few limited models, though speculators, such as Bill Gates, have suggested that they could become more common in the future. Many domestic robots are used for basic household chores. Others are educational or entertainment robots, such as the HERO line of the 1980s. While most domestic robots are simplistic, some are connected to WiFi home networks or smart environments and are autonomous to a high degree. There were an estimated 3,540,000 service robots in use in 2006, compared with an estimated 950,000 industrial robots.

Fig. : First generation Roomba vacuums the carpets in a domestic environment.

In-home Robots

This type of domestic robot does chores around and inside homes. Different kinds include:

Robotic vacuum cleaners and floor-washing robots that clean floors with sweeping and wet mopping functions. Some use Swiffer or other disposable cleaning cloths to dry-sweep, or reusable microfiber cloths to wet-mop.

Cat litter robots are automatic self-cleaning litter boxes that filter clumps out into a built-in waste receptacle that can be lined with an ordinary plastic bag.

Security robots which have a night-vision-capable wide-angle camera that detects movements and intruders. It can patrol places and shoot video of suspicious activities, too, and send alerts *via* email or text message; the stored history of past alerts and videos are accessible *via* the Web. The robot can also be configured to go into action at any time of the day.

Outdoor Robots

Outdoor robots are domestic robots that perform different chores that exist outside of the house. Robotic lawn mowers are one type of outdoor robot that cut grass on their own without the need for a driver. Some models can mow complicated and uneven lawns that are up to three-quarters of an acre in size. Others can mow a lawn as large as 40,000 square feet, can handle a hill inclined up to 27 degrees.

There are also automated pool cleaners that clean and maintain swimming pools autonomously by scrubbing in-ground pools from the floor to the waterline in 3 hours, cleaning and circulating more than 70 gallons of water per minute, and removing debris as small as 2 microns (2 millionths of a meter) in size.

Window-washing robots use two magnetic modules to navigate your windows as it sprays cleaning solution onto microfiber pads to wash them. It covers about 1601 square feet per charge. Gutter-cleaning robots can blast through debris, clogs, and sludge in gutters, and brushes them clean.

Toys

Robotic toys, such as the well known Furby, have been popular since 1998. There are also small humanoid remote controlled robots. Electronic pets, such as robotic dogs, can be companions for children. They have also have been used by many universities in competitions such as the RoboCup.

There are also phone-powered robots for fun and games, such as Romo which is a small robot that employs smartphones as its brain. By using another mobile device and a cross-platform app, the user can drive it, make it produce animated facial expressions, direct it to dance, or turn it into a spybot.

Social Robots

A social robot are robots whose main objective is social interaction. A lot of these robots are designed to help the elderly. For example, the Wakamaru is a humanoid robot designed to provide company for elderly and less mobile people, made by Mitsubishi Heavy Industries. There is also the Paro, a robot baby seal intended to provide comfort to nursing home patients.

Home-telepresence robots can move around in a remote location and let you communicate with people there *via* its camera, speaker, and microphone. Through other remote-controlled telepresence robots, anyone can visit a distant location and explore it as if they were physically present. It lets health-care workers check on patients easily, for instance, and kids who are homebound because of injuries, illnesses, or physical challenges can go to school *via* their VGo.

Robots no Longer in Production

Early historical attempts to bring robots into the home. Their actual functions is up for debate.

- Hero, 1982–89
- Topo, 1983.

In Popular Culture

Many cartoons feature robot maids, notably Rosie the Robot from *The Jetsons*. Maid Robots are especially prominent in anime (in Japanese, they are called Meido

Robo or Meido Roboto), and their Artificial Intelligence ranges from rudimentary to fully sentient and emotional, while their appearance ranges from obviously mechanical to human-like.

The 2012 movie *Robot & Frank* featured a domestic robot, the story of the movie centred around an elderly man and his relationship with a caretaker robot.

The 2005 movie *Robots* is a children's film, about a young inventor in a robot world traveling the big city to join his inspiration's company.

HUMANOID ROBOT

A humanoid robot is a robot with its body shape built to resemble that of the human body. A humanoid design might be for functional purposes, such as interacting with human tools and environments, for experimental purposes, such as the study of bipedal locomotion, or for other purposes. In general, humanoid robots have a torso, a head, two arms, and two legs, though some forms of humanoid robots may model only part of the body, for example, from the waist up. Some humanoid robots may also have heads designed to replicate human facial features such as eyes and mouths. Androids are humanoid robots built to aesthetically resemble humans.

Purpose

Fig. : TOPIO, a humanoid robot, playedping pong at Tokyo International Robot Exhibition (IREX) 2009.

Fig. : Nao (robot) is a robot created for companionship. It also competes in the RoboCupsoccer championship.

Humanoid robots are used as a research tool in several scientific area.

Researchers need to understand the human body structure and behavior (biomechanics) to build and study humanoid robots. On the other side, the attempt to the simulation of the human body leads to a better understanding of it.

Human cognition is a field of study which is focused on how humans learn from sensory information in order to acquire perceptual and motor skills. This knowledge is used to develop computational models of human behavior and it has been improving over time.

It has been suggested that very advanced robotics will facilitate the enhancement of ordinary humans.

Although the initial aim of humanoid research was to build better orthosis and prosthesis for human beings, knowledge has been transferred between both disciplines. A few examples are: powered leg prosthesis for neuromuscularly impaired, ankle-foot orthosis, biological realistic leg prosthesis and forearm prosthesis.

Besides the research, humanoid robots are being developed to perform human tasks like personal assistance, where they should be able to assist the sick and elderly, and dirty or dangerous jobs. Regular jobs like being a receptionist or a worker of an automotive manufacturing line are also suitable for humanoids. In essence, since they can use tools and operate equipment and vehicles designed for the human form, humanoids could theoretically perform any task a human being can, so long as they have the proper software. However, the complexity of doing so is deceptively great.

They are becoming increasingly popular for providing entertainment too. For example, Ursula, a female robot, sings, play music, dances, and speaks to her audiences at Universal Studios. Several Disney attractions employ the use of animatrons, robots that look, move, and speak much like human beings, in some of their theme park shows. These animatrons look so realistic that it can be hard to decipher from a distance whether or not they are actually human. Although they have a realistic look, they have no cognition or physical autonomy. Various humanoid robots and their possible applications in daily life are featured in an independent documentary film called *Plug & Pray*, which was released in 2010.

Humanoid robots, especially with artificial intelligence algorithms, could be useful for future dangerous and/or distant space explorationmissions, without having the need to turn back around again and return to Earth once the mission is completed.

Sensors

A sensor is a device that measures some attribute of the world. Being one of the three primitives of robotics (besides planning and control), sensing plays an important role in robotic paradigms. Sensors can be classified according to the physical process with which they work or according to the type of measurement information that they give as output. In this case, the second approach was used.

Fig. : <u>Enon</u> was created to be a personal assistant. It is self-guiding and has limited speech recognition and synthesis. It can also carry things.

Proprioceptive Sensors

Proprioceptive sensors sense the position, the orientation and the speed of the humanoid's body and joints. In human beings inner ears are used to maintain balance and orientation. Humanoid robots use accelerometers to measure the acceleration, from which velocity can be calculated by integration; tilt sensors to measure inclination; force sensors placed in robot's hands and feet to measure contact force with environment; position sensors, that indicate the actual position of the robot (from which the velocity can be calculated by derivation) or even speed sensors.

Exteroceptive Sensors

Fig. : An artificial hand holding a lightbulb.

Arrays of tactels can be used to provide data on what has been touched. The Shadow Hand uses an array of 34 tactels arranged beneath its polyurethane skin on each finger tip. Tactile sensors also provide information about forces and torques transferred between the robot and other objects. Vision refers to processing data from any modality which uses the electromagnetic spectrum to produce an image. In humanoid robots it is used to recognize objects and determine their properties. Vision sensors work most similarly to the eyes of human beings. Most humanoid robots use CCD cameras as vision sensors.

Sound sensors allow humanoid robots to hear speech and environmental sounds, and perform as the ears of the human being.Microphones are usually used for this task.

Actuators

Actuators are the motors responsible for motion in the robot. Humanoid robots are constructed in such a way that they mimic the human body, so they use actuators that perform like muscles andjoints, though with a different structure. To achieve the same effect as human motion, humanoid robots use mainly rotary actuators. They can be either electric, pneumatic, hydraulic, piezoelectric or ultrasonic.

Hydraulic and electric actuators have a very rigid behavior and can only be made to act in a compliant manner through the use of relatively complex feedback control strategies. While electric coreless motor actuators are better suited for high speed and low load applications, hydraulic ones operate well at low speed and high load applications. Piezoelectric actuators generate a small movement with a high force capability when voltage is applied. They can be used for ultra-precise positioning and for generating and handling high forces or pressures in static or dynamic situations.

Ultrasonic actuators are designed to produce movements in a micrometer order at ultrasonic frequencies (over 20 kHz). They are useful for controlling vibration, positioning applications and quick switching. Pneumatic actuators operate on the basis of gas compressibility. As they are inflated, they expand along the axis, and as they deflate, they contract. If one end is fixed, the other will move in a linear trajectory. These actuators are intended for low speed and low/medium load applications. Between pneumatic actuators there are: cylinders,bellows, pneumatic engines, pneumatic stepper motors and pneumatic artificial muscles.

Planning and Control

In planning and control, the essential difference between humanoids and other kinds of robots (like industrial ones) is that the movement of the robot has to be human-like, using legged locomotion, especially biped gait. The ideal planning for humanoid movements during normal walking should result in minimum energy consumption, like it does in the human body. For this reason, studies on dynamics and control of these kinds of structures become more and more important. To maintain dynamic balance during the walk, a robot needs information about contact force and its current and desired motion. The solution to this problem relies on a major concept, the Zero Moment Point (ZMP).

Another characteristic of humanoid robots is that they move, gather information (using sensors) on the "real world" and interact with it. They don't stay still like factory manipulators and other robots that work in highly structured environments. To allow humanoids to move in complex environments, planning and control must focus on self-collision detection, path planning and obstacle avoidance.

Humanoids don't yet have some features of the human body. They include structures with variable flexibility, which provide safety (to the robot itself and to the people), and redundancy of movements, *i.e.* more degrees of freedom and therefore wide task availability. Although these characteristics are desirable to humanoid robots, they will bring more complexity and new problems to planning and control.

Timeline of Developments

Year	Development
c. 250 BC	The *Lie Zi* described an automaton.
c. 50 AD	Greek mathematician Hero of Alexandria described a machine to automatically pour wine for party guests.
1206	Al-Jazari described a band made up of humanoid automata which, according to Charles B. Fowler, performed "more than fifty facial and body actions during each musical selection." Al-Jazari also created hand washing automata with automatic humanoid servants, and an elephant clock incorporating an automatic humanoid mahout striking a cymbal on the half-hour. His programmable "castle clock" also featured five musician automata which automatically played music when moved by levers operated by a hidden camshaft attached to a water wheel.
1495	Leonardo da Vinci designs a humanoid automaton that looks like an armored knight, known as Leonardo's robot.
1738	Jacques de Vaucanson builds The Flute Player, a life-size figure of a shepherd that could play twelve songs on the flute and The Tambourine Player that played a flute and a drum or tambourine.
1774	Pierre Jacquet-Droz and his son Henri-Louis created the Draughtsman, the Musicienne and the Writer, a figure of a boy that could write messages up to 40 characters long.
1898	Nikola Tesla publicly demonstrates his "automaton" technology by wirelessly controlling a model boat at the Electrical Exposition held at Madison Square Garden in New York City during the height of the Spanish-American War.
1921	Czech writer Karel Čapek introduced the word "robot" in his play *R.U.R. (Rossum's Universal Robots)*. The word "robot" comes from the word "robota", meaning, in Czech and Polish, "forced labour, drudgery".
1927	The Maschinenmensch ("machine-human"), a gynoid humanoid robot, also called "Parody", "Futura", "Robotrix", or the "Maria impersonator" (played by German actress Brigitte Helm), perhaps the most memorable humanoid robot ever to appear on film, is depicted in Fritz Lang's film Metropolis.
1941-42	Isaac Asimov formulates the Three Laws of Robotics, and in the process of doing so, coins the word "robotics".
1948	Norbert Wiener formulates the principles of cybernetics, the basis of practical robotics.

1961	The first digitally operated and programmable non-humanoid robot, the Unimate, is installed on a General Motors assembly line to lift hot pieces of metal from a die casting machine and stack them. It was created by George Devol and constructed by Unimation, the first robot manufacturing company.
1969	D.E. Whitney publishes his article "Resolved motion rate control of manipulators and human prosthesis".
1970	Miomir Vukobratović has proposed Zero Moment Point, a theoretical model to explain biped locomotion.
1972	Miomir Vukobratović and his associates at Mihajlo Pupin Institute build the first active anthropomorphic exoskeleton.
1973	In Waseda University, in Tokyo, Wabot-1 is built. It was able to walk, to communicate with a person in Japanese and to measure distances and directions to the objects using external receptors, artificial ears and eyes, and an artificial mouth.
1980	Marc Raibert established the MIT Leg Lab, which is dedicated to studying legged locomotion and building dynamic legged robots.
1983	Using MB Associates arms, "Greenman" was developed by Space and Naval Warfare Systems Center, San Diego. It had an exoskeletal master controller with kinematic equivalency and spatial correspondence of the torso, arms, and head. Its vision system consisted of two 525-line video cameras each having a 35-degree field of view and video camera eyepiece monitors mounted in an aviator's helmet.
1984	At Waseda University, the Wabot-2 is created, a musician humanoid robot able to communicate with a person, read a normal musical score with his eyes and play tunes of average difficulty on an electronic organ.
1985	Developed by Hitachi Ltd, WHL-11 is a biped robot capable of static walking on a flat surface at 13 seconds per step and it can also turn.
1985	WASUBOT is another musician robot from Waseda University. It performed a concerto with the NHK Symphony Orchestra at the opening ceremony of the International Science and Technology Exposition.
1986	Honda developed seven biped robots which were designated E0 (Experimental Model 0) through E6. E0 was in 1986, E1 – E3 were done between 1987 and 1991, and E4 - E6 were done between 1991 and 1993.
1989	Manny was a full-scale anthropomorphic robot with 42 degrees of freedom developed at Battelle's Pacific Northwest Laboratories in Richland, Washington, for the US Army's Dugway Proving Ground in Utah. It could not walk on its own but it could crawl, and had an artificial respiratory system to simulate breathing and sweating.
1990	Tad McGeer showed that a biped mechanical structure with knees could walk passively down a sloping surface.
1993	Honda developed P1 (Prototype Model 1) through P3, an evolution from E series, with upper limbs. Developed until 1997.
1995	Hadaly was developed in Waseda University to study human-robot communication and has three subsystems: a head-eye subsystem, a voice control system for listening and speaking in Japanese, and a motion-control subsystem to use the arms to point toward campus destinations.

1995	Wabian is a human-size biped walking robot from Waseda University.
1996	Saika, a light-weight, human-size and low-cost humanoid robot, was developed at Tokyo University. Saika has a two-DOF neck, dual five-DOF upper arms, a torso and a head. Several types of hands and forearms are under development also. Developed until 1998.
1997	Hadaly-2, developed at Waseda University, is a humanoid robot which realizes interactive communication with humans. It communicates not only informationally, but also physically.
2000	Honda creates its 11th bipedal humanoid robot, able to run, ASIMO.
2001	Sony unveils small humanoid entertainment robots, dubbed Sony Dream Robot (SDR). Renamed Qrio in 2003.
2001	Fujitsu realized its first commercial humanoid robot named HOAP-1. Its successors HOAP-2 and HOAP-3 were announced in 2003 and 2005, respectively. HOAP is designed for a broad range of applications for R&D of robot technologies.
2002	HRP-2, biped walking robot built by the Manufacturing Science and Technology Center (MSTC) in Tokyo.
2003	JOHNNIE, an autonomous biped walking robot built at the Technical University of Munich. The main objective was to realize an anthropomorphic walking machine with a human-like, dynamically stable gait
2003	Actroid, a robot with realistic silicone "skin" developed by Osaka University in conjunction with Kokoro Company Ltd.
2004	Persia, Iran's first humanoid robot, was developed using realistic simulation by researchers of Isfahan University of Technology in conjunction with ISTT.
2004	KHR-1, a programmable bipedal humanoid robot introduced in June 2004 by a Japanese company Kondo Kagaku.
2005	The PKD Android, a conversational humanoid robot made in the likeness of science fiction novelist Philip K Dick, was developed as a collaboration between Hanson Robotics, the FedEx Institute of Technology, and the University of Memphis.
2005	Wakamaru, a Japanese domestic robot made by Mitsubishi Heavy Industries, primarily intended to provide companionship to elderly and disabled people.
2005	Nao is a small open source programmable humanoid robot developed by Aldebaran Robotics, in France. Widely used by world wide universities as a research platform and educational tool.
2005	The Geminoid series is a series of ultra-realistic humanoid robots or Actroid developed by Hiroshi Ishiguro of ATR and Kokoro in Tokyo. The original one, Geminoid HI-1 was made at its image. Followed Geminoid-F in 2010 and Geminoid-DK in 2011.
2006	RoboTurk is designed and realized by Dr Davut Akdas and Dr Sabri Bicakci at Balikesir University. This Research Project Sponsored By The Scientific And Technological Research Council Of Turkey (TUBITAK) in 2006. RoboTurk is successor of biped robots named "Salford Lady" and "Gonzalez" at university of Salford in the UK. It is the first humanoid robot supported by Turkish Government.
2006	REEM-A, a biped humanoid robot designed to play chess with the Hydra Chess engine. The first robot developed by PAL Robotics, it was also used as a walking, manipulation speech and vision development platform.

2006	iCub, a biped humanoid open source robot for cognition research.
2006	Mahru, a network-based biped humanoid robot developed in South Korea.
2007	TOPIO, a ping pong playing robot developed by TOSY Robotics JSC.
2007	Twendy-One, a robot developed by the WASEDA University Sugano Laboratory for home assistance services. It is not biped, as it uses an omni-directional mobile mechanism.
2008	Justin, a humanoid robot developed by the German Aerospace Center (DLR).
2008	KT-X, the first international humanoid robot developed as a collaboration between the five-time consecutive RoboCup champions, Team Osaka, and KumoTek Robotics.
2008	Nexi, the first mobile, dexterous and social robot, makes its public debut as one of *TIME* magazine's top inventions of the year. The robot was built through a collaboration between the MIT Media Lab Personal Robots Group, UMass Amherst and Meka robotics.
2008	Salvius (robot), The first open source humanoid robot built in the United States is created.
2008	REEM-B, the second biped humanoid robot developed by PAL Robotics. It has the ability to autonomously learn its environment using various sensors and carry 20% of its own weight.
2009	HRP-4C, a Japanese domestic robot made by National Institute of Advanced Industrial Science and Technology, shows human characteristics in addition to bipedal walking.
2009	Turkey's first dynamically walking humanoid robot, SURALP, is developed by Sabanci University in conjunction with Tubitak.
2009	Kobian, an robot developed by WASEDA University can walk, talk and mimic emotions.
2010	NASA and General Motors revealed Robonaut 2, a very advanced humanoid robot. It was part of the payload of Shuttle Discovery on the successful launch February 24, 2011. It is intended to do spacewalks for NASA.
2010	Students at the University of Tehran, Iran unveil the Surena II. It was unveiled by President Mahmoud Ahmadinejad.
2010	Researchers at Japan's National Institute of Advanced Industrial Science and Technology demonstrate their humanoid robot HRP-4C singing and dancing along with human dancers.
2010	In September the National Institute of Advanced Industrial Science and Technology also demonstrates the humanoid robot HRP-4. The HRP-4 resembles the HRP-4C in some regards but is called "athletic" and is not a gynoid.
2010	REEM, a humanoid service robot with a wheeled mobile base. Developed by PAL Robotics, it can perform autonomous navigation in various surroundings and has voice and face recognition capabilities.
2011	In November Honda unveiled its second generation Honda Asimo Robot. The all new Asimo is the first version of the robot with semi-autonomous capabilities.
2012	In April, the Advanced Robotics Department in Italian Institute of Technology released its first version of the COmpliant huMANoid robot COMAN which is designed for robust dynamic walking and balancing in rough terrain.

HEXAPOD (ROBOTICS)

Fig. : A desktop sized hexapod.

A hexapod robot is a mechanical vehicle that walks on six legs. Since a robot can be statically stable on three or more legs, a hexapod robot has a great deal of flexibility in how it can move. If legs become disabled, the robot may still be able to walk. Furthermore, not all of the robot's legs are needed for stability; other legs are free to reach new foot placements or manipulate a payload.

Many hexapod robots are biologically inspired by Hexapoda locomotion. Hexapods may be used to test biological theories about insect locomotion, motor control, and neurobiology.

Designs

Fig. : Two hexapod robots at the Georgia Institute of Technology with CMUCams mounted on top.

Hexapod designs vary in leg arrangement. Insect-inspired robots are typically laterally symmetric, such as the RiSE robot at Carnegie Mellon. A radially symmetric hexapod is ATHLETE (All-Terrain Hex-Legged Extra-Terrestrial Explorer) robot at JPL.

Typically, individual legs range from two to six degrees of freedom. Hexapod feet are typically pointed, but can also be tipped with adhesive material to help climb walls or wheels so the robot can drive quickly when the ground is flat.

Locomotion

Most often, hexapods are controlled by gaits, which allow the robot to move forward, turn, and perhaps side-step. Some of the most common gaits are as follows:

- Alternating tripod: 3 legs on the ground at a time.
- Quadruped.
- Crawl: move just one leg at a time.

Gaits for hexapods are often stable, even in slightly rocky and uneven terrain.

Motion may also be nongaited, which means the sequence of leg motions is not fixed, but rather chosen by the computer in response to the sensed environment. This may be most helpful in very rocky terrain, but existing techniques for motion planning are computationally expensive.

Biologically Inspired

Insects are chosen as models because their nervous system is simpler than other animal species. Also, complex behaviours can be attributed to just a few neurons and the pathway between sensory input and motor output is relatively shorter. Insects' walking behaviour and neural architecture are used to improve robot locomotion. Alternatively, biologists can use hexapod robots for testing different hypotheses.

Biologically inspired hexapod robots largely depend on the insect species used as a model. The cockroach and the stick insect are the two most commonly used insect species; both have been ethologically and neurophysiologically extensively studied. At present no complete nervous system is known, therefore, models usually combine different insect models, including those of other insects.

Insect gaits are usually obtained by two approaches: the centralized and the decentralized control architectures. Centralized controllers directly specify transitions of all legs, whereas in decentralized architectures, six nodes (legs) are connected in a parallel network; gaits arise by the interaction between neighbouring legs.

LEG SENSOR INTEGRATION FOR HEXAPOD ROBOT MOTION

This project is a collaboration of many departments on this campus, such as Bioengineering, Electrical and Computer Engineering, Mechanical Engineering, Computer Science, and Neuroscience. This project will be very educational because a lot of professionals are doing research in different fields, collecting the data and eventually transferring it to the Beckman Institute where the physical robot is being built. My main task will be improving the leg motion of the robot by integration of sensors. This sounds like a very exciting project.

Goals and Intentions

The main goals of my project are leg sensor interfacing, software modification to robot controller, sensor data acquisition, and acquiring 'normal' readings from sensors during leg motions, and provide behavioral responses to sensor information. Leg sensor interfacing includes testing of surface mount board, rewiring incorrect traces, and testing new sensors with old sensor algorithm. Software modifications to robot controller includes sampling sensor data, creating leg motion, and using data to effect leg motion. Sensor data acquisition includes collecting files of data to analyze, looking at data in Matlab, and adjusting the gains on individual sensors. Taking normal readings from sensors during leg motions includes mapping out normal readings during leg motions and looking at situations that create abnormal readings. Providing behavioral reactions to sensor information includes providing reactions such as swinging the leg higher and re-stepping in response to the abnormal readings.

Block Functions

The block diagram consists of sensors that read the data according to the environment and sends to the software which in turn detects the kind of actions that need to take based on the acquired data. The signals from the software get sent to the actuators where the actual physical motion takes place.

There are four types of sensors: foot sensors, strain gages, pressure sensors, and position sensors. The foot sensors detect the ground conditions with the 'Toe' and also detect forces at the 'Toe' to help with the thrusting motion. The strain gages measure load on a leg segments to balance the weight of the robot. The pressure sensors monitor air pressure in each of the cylinder chambers to measure the force it is generating. The position sensors measure angle between two adjacent leg segments so walking motions can be generated and maintained.

The software reads input from the sensors, decides on leg motion according to the readings, and produces valve signals to move the legs. The software then generates signals, which are sent to the flexor and extensor air valves, to perform the appropriate leg motions.

Performance of the Finished Project

At the end of the project, the requirements for the performance are an adequate amplification of the sensors to be read by the computer input/output (I/O) board and the sensor data must be able to detect different situations the leg will encounter. For example, some sensors have very low output voltages that must be adjusted to a range between 0-5V for the I/O board to be able to read them. Also, if the leg hits something during a swing, or if it touches the ground, the sensors must be able to detect that these situations and then the software must be able to detect which situation has occurred.

INDUSTRIAL ROBOT

Fig. : Articulated industrial robot operating in a foundry.

An industrial robot is defined by ISO as an *automatically controlled, reprogrammable, multipurpose manipulator programmable in three or more axes*. The field of robotics may be more practically defined as the study, design and use of robot systems formanufacturing (a top-level definition relying on the prior definition of *robot*).

Typical applications of robots include welding, painting, assembly, pick and place (such as packaging, palletizing and SMT), product inspection, and testing; all accomplished with high endurance, speed, and precision.

Robot Types, Features

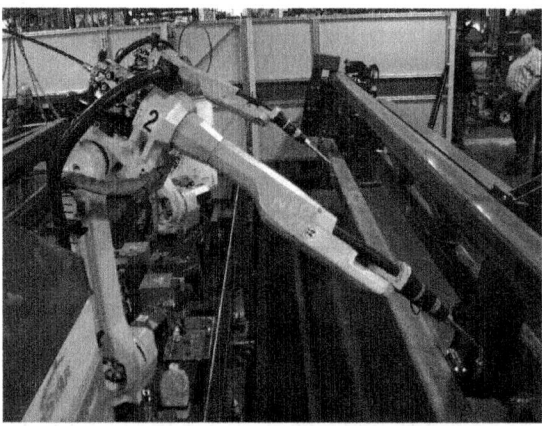

Fig. : A set of six-axis robots used forwelding.

Fig. : Factory Automation with industrial robots for palletizing food products like bread and toast at a bakery in Germany.

The most commonly used robot configurations are articulated robots, SCARA robots, Delta robots and Cartesian coordinate robots, (aka gantry robots or x-y-z robots). In the context of general robotics, most types of robots would fall into the category ofrobotic arms (inherent in the use of the word *manipulator* in the above-mentioned ISO standard). Robots exhibit varying degrees ofautonomy:

- Some robots are programmed to faithfully carry out specific actions over and over again (repetitive actions) without variation and with a high degree of accuracy. These actions are determined by programmed routines that specify the direction, acceleration, velocity, deceleration, and distance of a series of coordinated motions.

- Other robots are much more flexible as to the orientation of the object on which they are operating or even the task that has to be performed on the object itself, which the robot may even need to identify.

History of Industrial Robotics

The earliest known industrial robot, conforming to the ISO definition was completed by "Bill" Griffith P. Taylor in 1937 and published in Meccano Magazine, March 1938. The crane-like device was built almost entirely using Meccano parts, and powered by a single electric motor. Five axes of movement were possible, including Grab and Grab Rotation. Automation was achieved using punched paper tape to energise solenoids, which would facilitate the movement of the crane's control levers. The robot could stack wooden blocks in pre-programmed patterns. The number of motor revolutions required for each desired movement was first plotted on graph paper. This information was then transferred to the paper tape, which was also driven by the robot's single motor. Chris Shute built a complete replica of the robot in 1997.

George Devol applied for the first robotics patents in 1954 (granted in 1961). The first company to produce a robot was Unimation, founded by Devol and Joseph F. Engelberger in 1956, and was based on Devol's original patents. Unimation robots were also called *programmable transfer machines* since their main use at first was to transfer objects from one point to another, less than a dozen feet or so apart. They used hydraulic actuators and were programmed in *joint coordinates, i.e.* the angles of the various joints were stored during a teaching phase and replayed in operation. They were accurate to within 1/10,000 of an inch (note: although accuracy is not an appropriate measure for robots, usually evaluated in terms of repeatability). Unimation later licensed their technology to Kawasaki Heavy Industries and GKN, manufacturing Unimates in Japan and England respectively. For some time Unimation's only competitor was Cincinnati Milacron Inc. of Ohio. This changed radically in the late 1970s when several big Japanese conglomerates began producing similar industrial robots.

In 1969 Victor Scheinman at Stanford University invented the Stanford arm, an all-electric, 6-axis articulated robot designed to permit an arm solution. This allowed it accurately to follow arbitrary paths in space and widened the potential use of the robot to more sophisticated applications such as assembly and welding. Scheinman then designed a second arm for the MIT AI Lab, called the "MIT arm." Scheinman, after receiving a fellowship from Unimation to develop his designs, sold those designs to Unimation who further developed them with support from General Motors and later marketed it as the Programmable Universal Machine for Assembly (PUMA).

Industrial robotics took off quite quickly in Europe, with both ABB Robotics and KUKA Robotics bringing robots to the market in 1973. ABB Robotics (formerly ASEA) introduced IRB 6, among the world's first *commercially available* all electric micro-processor controlled robot. The first two IRB 6 robots were sold to Magnusson in Sweden for grinding and polishing pipe bends and were installed in production in January 1974. Also in 1973 KUKA Robotics built its first robot, known as FAMULUS, also one of the first articulated robots to have six electro-mechanically driven axes.

Interest in robotics increased in the late 1970s and many US companies entered the field, including large firms like General Electric, and General Motors (which formed joint venture FANUC Robotics with FANUC LTD of Japan). U.S. startup companies included Automatix and Adept Technology, Inc. At the height of the robot boom in 1984, Unimation was acquired by Westinghouse Electric Corporation for 107 million U.S. dollars. Westinghouse sold Unimation to Stäubli Faverges SCA of France in 1988, which is still making articulated robots for general industrial and cleanroom applications and even bought the robotic division of Bosch in late 2004.

Only a few non-Japanese companies ultimately managed to survive in this market, the major ones being: Adept Technology, Stäubli-Unimation, the Swedish-Swiss company ABB Asea Brown Boveri, the German company KUKA Robotics and the Italian company Comau.

TECHNICAL DESCRIPTION

Defining Parameters

- **Number of axes** – two axes are required to reach any point in a plane; three axes are required to reach any point in space. To fully control the orientation of the end of the arm (*i.e.* the wrist) three more axes (yaw, pitch, and roll) are required. Some designs (*e.g.* the SCARA robot) trade limitations in motion possibilities for cost, speed, and accuracy.
- Degrees of freedom which is usually the same as the number of axes.
- **Working envelope** – the region of space a robot can reach.
- **Kinematics** – the actual arrangement of rigid members and joints in the robot, which determines the robot's possible motions. Classes of robot kinematics include articulated, cartesian, parallel and SCARA.
- **Carrying capacity or payload** – how much weight a robot can lift.
- **Speed** – how fast the robot can position the end of its arm. This may be defined in terms of the angular or linear speed of each axis or as a compound speed *i.e.* the speed of the end of the arm when all axes are moving.
- **Acceleration** - how quickly an axis can accelerate. Since this is a limiting factor a robot may not be able to reach its specified maximum speed for movements over a short distance or a complex path requiring frequent changes of direction.
- **Accuracy** – how closely a robot can reach a commanded position. When the absolute position of the robot is measured and compared to the commanded position the error is a measure of accuracy. Accuracy can be improved with external sensing for example a vision system or Infra-Red. Accuracy can vary with speed and position within the working envelope and with payload.
- **Repeatability** - how well the robot will return to a programmed position. This is not the same as accuracy. It may be that when told to go to a certain X-Y-Z position that it gets only to within 1 mm of that position. This would be its accuracy which may be improved by calibration. But if that position is taught into controller memory and each time it is sent there it returns to within 0.1 mm of the taught position then the repeatability will be within 0.1 mm.

Accuracy and repeatability are different measures. Repeatability is usually the most important criterion for a robot and is similar to the concept of 'precision' in measurement ISO 9283 sets out a method whereby both accuracy and repeatability can be measured. Typically a robot is sent to a taught position a number of times and the error is measured at each return to the position after visiting 4 other positions. Repeatability is then quantified using the standard deviation of those samples in all three dimensions. A typical robot can, of course make a positional error exceeding that and that could be a problem for the process. Moreover the repeatability is different in different parts of the working envelope and also changes with speed and payload. ISO 9283 specifies that accuracy and repeatability should be measured at maximum speed and at maximum payload. But this results in pessimistic values whereas the robot could be much more accurate and repeatable at light loads and speeds. Repeatability in an industrial process is also subject to the accuracy of the end effector, for example a gripper, and even to the design of the 'fingers' that match the gripper to the object being grasped. For example if a robot picks a screw by its head the screw could be at a random angle. A subsequent attempt to insert the screw into a hole could easily fail. These and similar scenarios can be improved with 'lead-ins' *e.g.* by making the entrance to the hole tapered.

- **Motion control** – for some applications, such as simple pick-and-place assembly, the robot need merely return repeatably to a limited number of pre-taught positions.

- **Power source** – some robots use electric motors, others use hydraulic actuators. The former are faster, the latter are stronger and advantageous in applications such as spray painting, where a spark could set off an explosion; however, low internal air-pressurisation of the arm can prevent ingress of flammable vapours as well as other contaminants.

- **Drive** – some robots connect electric motors to the joints *via* gears; others connect the motor to the joint directly (direct drive). Using gears results in measurable 'backlash' which is free movement in an axis. Smaller robot arms frequently employ high speed, low torque DC motors, which generally require high gearing ratios; this has the disadvantage of backlash. In such cases the harmonic drive is often used.

- **Compliance** - this is a measure of the amount in angle or distance that a robot axis will move when a force is applied to it. Because of compliance when a robot goes to a position carrying its maximum payload it will be at a position slightly lower than when it is carrying no payload. Compliance can also be responsible for overshoot when carrying high payloads in which case acceleration would need to be reduced.

Robot Programming and Interfaces

The setup or programming of motions and sequences for an industrial robot is typically taught by linking the robot controller to alaptop, desktop computer or (internal or Internet) network. A robot and a collection of machines or peripherals is referred to as a workcell, or cell. A typical cell might contain a parts feeder, amolding machine and a robot. The various machines are 'integrated' and controlled by a single computer or PLC. How the robot interacts with other machines in the cell must be programmed, both with regard to their positions in the cell and synchronizing with them.

Software: The computer is installed with corresponding interface software. The use of a computer greatly simplifies the programming process. Specialized robot software is run either in the robot controller or in the computer or both depending on the system design.

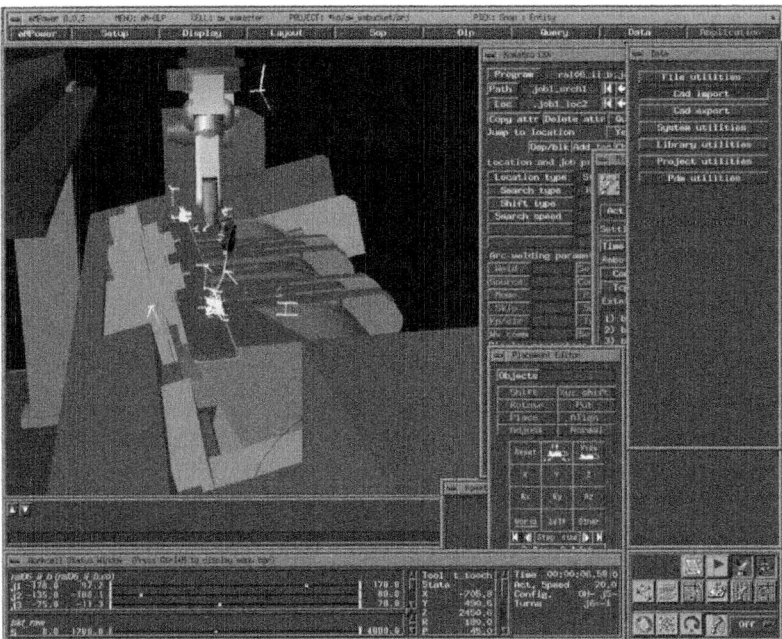

Fig. : Offline programming by ROBCAD.

Fig. : A typical well-used teach pendant with optional mouse.

There are two basic entities that need to be taught (or programmed): positional data and procedure. For example in a task to move a screw from a feeder to a hole the positions of the feeder and the hole must first be taught or programmed. Secondly the procedure to get the screw from the feeder to the hole must be programmed along with any I/O involved, for example a signal to indicate when the screw is in the feeder ready to be picked up. The purpose of the robot software is to facilitate both these programming tasks.

Teaching the robot positions may be achieved a number of ways:

Positional commands The robot can be directed to the required position using a GUI or text based commands in which the required X-Y-Z position may be specified and edited.

Teach pendant: Robot positions can be taught *via* a teach pendant. This is a handheld control and programming unit. The common features of such units are the ability to manually send the robot to a desired position, or "inch" or "jog" to adjust a position. They also have a means to change the speed since a low speed is usually required for careful positioning, or while test-running through a new or modified routine. A large emergency stop button is usually included as well. Typically once the robot has been programmed there is no more use for the teach pendant.

Lead-by-the-nose is a technique offered by many robot manufacturers. In this method, one user holds the robot's manipulator, while another person enters a command which de-energizes the robot causing it to go limp. The user then moves the robot by hand to the required positions and/or along a required path while the software logs these positions into memory. The program can later run the robot to these positions or along the taught path. This technique is popular for tasks such as paint spraying.

Offline programming is where the entire cell, the robot and all the machines or instruments in the workspace are mapped graphically. The robot can then be moved on screen and the process simulated. The technique has limited value because it relies on accurate measurement of the positions of the associated equipment and also relies on the positional accuracy the robot which may or may not conform to what is programmed.

Others In addition, machine operators often use user interface devices, typically touchscreen units, which serve as the operator control panel. The operator can switch from program to program, make adjustments within a program and also operate a host of peripheral devices that may be integrated within the same robotic system. These includeend effectors, feeders that supply components to the robot, conveyor belts, emergency stop controls, machine vision systems, safety interlock systems, bar code printers and an almost infinite array of other industrial devices which are accessed and controlled *via* the operator control panel.

The teach pendant or PC is usually disconnected after programming and the robot then runs on the program that has been installed in its controller. However a computer is often used to 'supervise' the robot and any peripherals, or to provide additional storage for access to numerous complex paths and routines.

End-of-arm Tooling

The most essential robot peripheral is the end effector, or end-of-arm-tooling (EOT). Common examples of end effectors include welding devices (such as MIG-welding guns, spot-welders, *etc.*), spray guns and also grinding and deburring devices (such as pneumatic disk or belt grinders, burrs, *etc.*), and grippers (devices that can grasp an object, usually electromechanical or pneumatic). Another common means of picking up an object is by vacuum. End effectors are frequently highly complex, made to match the handled product and often capable of picking up an array of products at one time. They may utilize various sensors to aid the robot system in locating, handling, and positioning products.

Controlling Movement

For a given robot the only parameters necessary to completely locate the end effector (gripper, welding torch, *etc.*) of the robot are the angles of each of the joints or displacements of the linear axes (or combinations of the two for robot formats such as SCARA). However there are many different ways to define the points. The most common and most convenient way of defining a point is to specify a Cartesian coordinate for it, *i.e.* the position of the 'end effector' in mm in the X, Y and Z directions relative to the robot's origin. In addition, depending on the types of joints a particular robot may have, the orientation of the end effector in yaw, pitch, and roll and the location of the tool point relative to the robot's faceplate must also be specified. For a jointed arm these coordinates must be converted to joint angles by the robot controller and such conversions are known as Cartesian Transformations which may need to be performed iteratively or recursively for a multiple axis robot. The mathematics of the relationship between joint angles and actual spatial coordinates is called kinematics.

Positioning by Cartesian coordinates may be done by entering the coordinates into the system or by using a teach pendant which moves the robot in X-Y-Z directions. It is much easier for a human operator to visualize motions up/down, left/right, *etc.* than to move each joint one at a time. When the desired position is reached it is then defined in some way particular to the robot software in use, *e.g.* P1 - P5 below.

Typical Programming

Most articulated robots perform by storing a series of positions in memory, and moving to them at various times in their programming sequence. For example, a robot which is moving items from one place to another might have a simple 'pick and place' program similar to the following:

Define points P1–P5:

1. Safely above workpiece (defined as P1)
2. 10 cm Above bin A (defined as P2)
3. At position to take part from bin A (defined as P3)

4. 10 cm Above bin B (defined as P4)

5. At position to take part from bin B. (defined as P5)

Define program:

1. Move to P1

2. Move to P2

3. Move to P3

4. Close gripper

5. Move to P2

6. Move to P4

7. Move to P5

8. Open gripper

9. Move to P4

10. Move to P1 and finish

Singularities

The American National Standard for Industrial Robots and Robot Systems — Safety Requirements (ANSI/RIA R15.06-1999) defines a singularity as "a condition caused by the collinear alignment of two or more robot axes resulting in unpredictable robot motion and velocities." It is most common in robot arms that utilize a "triple-roll wrist". This is a wrist about which the three axes of the wrist, controlling yaw, pitch, and roll, all pass through a common point. An example of a wrist singularity is when the path through which the robot is traveling causes the first and third axes of the robot's wrist (*i.e.* robot's axes 4 and 6) to line up.

The second wrist axis then attempts to spin 360° in zero time to maintain the orientation of the end effector. Another common term for this singularity is a "wrist flip". The result of a singularity can be quite dramatic and can have adverse effects on the robot arm, the end effector, and the process. Some industrial robot manufacturers have attempted to side-step the situation by slightly altering the robot's path to prevent this condition. Another method is to slow the robot's travel speed, thus reducing the speed required for the wrist to make the transition. The ANSI/RIA has mandated that robot manufacturers shall make the user aware of singularities if they occur while the system is being manually manipulated.

A second type of singularity in wrist-partitioned vertically articulated six-axis robots occurs when the wrist center lies on a cylinder that is centered about axis 1 and with radius equal to the distance between axes 1 and 4. This is called a shoulder singularity. Some robot manufacturers also mention alignment singularities, where axes 1 and 6 become coincident. This is simply a sub-case of shoulder singularities. When the robot passes close to a shoulder singularity, joint 1 spins very fast.

The third and last type of singularity in wrist-partitioned vertically articulated six-axis robots occurs when the wrist's center lies in the same plane as axes 2 and 3.

Singularities are closely related to the phenomena of Gimbal Lock, which has a similar root cause of axes becoming lined up.

A video illustrating these three types of singular configurations is available here.

Recent and Future Developments

As of 2005, the robotic arm business is approaching a mature state, where they can provide enough speed, accuracy and ease of use for most of the applications. Vision guidance (aka machine vision) is bringing a lot of flexibility to robotic cells. However, the end effector attached to a robot is often a simple pneumatic, 2-position chuck. This does not allow the robotic cell to easily handle different parts, in different orientations.

Hand-in-hand with increasing off-line programmed applications, robot calibration is becoming more and more important in order to guarantee a good positioning accuracy.

Other developments include downsizing industrial arms for light industrial use such as production of small products, sealing and dispensing, quality control, handling samples in the laboratory. Such robots are usually classified as "bench top" robots. Robots are used in pharmaceutical research in a technique called High-throughput screening. Bench top robots are also used in consumer applications (micro-robotic arms). Industrial arms may be used in combination with or even mounted on automated guided vehicles (AGVs) to make the automation chain more flexible between pick-up and drop-off.

MOBILE ROBOT SYSTEMS

Designing a mobile robot language requires knowledge of the robot's physical configuration and the desired complexity of the implementation. We will define the term "robot". From this definition requirements for a generalized control scheme will emerge. Finally, specific details of the LLAMA system pertaining to robot control will be explained. We have established a message passing requirement-control flows from high to low level while reports flow in the opposite direction. Such a distinction has some impact on the design of a language.

Definition of Autonomous Mobile Robots

A robot is *any device which can be physically isolated that can cause a change in its environment in response to sensory stimuli*. From this definition, we see that the following can be considered robots: manipulator with force-feedback sensors; miniature maze-traversing autonomous robots. The physical isolation criterion requires that the hardware of the robot can be moved (practicality aside) without upsetting the functionality of the robot.

An arbitrary complex system may have many of the features of a robot and not be a robot: A desk-top computer is not a robot whereas a computer that controls a mobile platform is a robot. The desk-top computer can be referred to as a *static tool*, whereas a robot could be classified as an *active tool*. The presumed robot must have the capability to move through its environment before it can be considered a robot. An automobile cannot be considered a robot since it is a static tool; if the automobile has anti-lock brakes, which sense and affect the environment, albeit slightly, it could be considered a robot.

In our society it is difficult to view an automobile as a robot-it really is a mass of metal and plastic hurtling a human occupant to some destination. An automated guide vehicle, like the Sea-Tac automated subway system can be considered a robot. Note that if a mobile platform is controlled by a computer physically removed from the platform, then the platform can be considered a robot and part of a robot system. This is an important observation since this very nearly describes the LLAMA and G2 mobile robot system.

Mobile Robots are a class of robots that have the capability to transport themselves in three dimensional space. As was just observed, the controller need not be physically present on the mobile platform. To describe this difference in configuration, we will define *Open Autonomous System* as a mobile robot that is controlled by an external system, but not by a person. We will define a Closed Autonomous System as a mobile robot that is entirely self contained. Using this nomenclature and extending it appropriately, we can describe the LLAMA and G2 project as an *Open Near-Autonomous System with capabilities for Closed Autonomy*. This says that LLAMA could be programmed to perform some behavior with no external control, although under most circumstances would be controlled by the G2 expert system. The ultimate goal in robotics is an *Autonomous System,* either open or closed.

An *autonomous agent* is an abstraction of an autonomous system. Determining whether a system is truly autonomous requires some formal metrics, which is beyond the scope of this discussion. Many models that represent autonomous (robot) agents separate the robot from the environment. Kaelbling presents one possible model of an autonomous system that has the capability for learning by noting that an "agent can be in many different states of information about the environment, and it must map each of these information states, or situations, to a particular action that it can perform in the world". The system she suggests is reactive, which appears to be an essential quality for an autonomous agent. Implementing reactive behavior is trivial; implementing learning is a very difficult problem and will not be covered except to note that the LLAMA and G2 systems contain resources that might be configured for learning.

The Hardware-Sensors and Effectors

An open mobile robotic system will have computers with software on and off board the robot. On-board refers to physically residing on the mobile platform, while off-board refers to a location that is independent of the platform. In general, the on-board system will interface with the robot platform hardware and the off-board system will interface with the on-board system. One exception to this generalization is the stationary video camera that connects to the off-board system to monitor and affect the platform's movements. For our purposes, only the hardware that can be accessed by the on-board system is of interest.

A mobile robot system that can analyze and react to its environment through the robot platform is called *robo-centric*. A robot interacts with the environment by changing its physical presence over time. The robot may change the environment when it enables an effector or senses the environment through its sensors. Its perception of the environment is constantly changing if the robot system senses in a robo- centric manner. For example, even though the robot may not move a great distance and may not change its environment significantly, it is very likely that all sensor readings will have changed their values.

This large-scale sensor update can be a great burden on sensor processing and interpreting systems. G2 attempts to dampen this potential speed bottle-neck by introducing an environmental modeler as part of its knowledge base structure. In essence the LLAMA and G2 system is robo-centric, but contains an environmental modeler to reduce the complexity of sensor processing. This modeler expects certain environmental qualities as being absolute, such as level floors, ever-present (read: detectable) walls, landmark recognition, *etc.* These environmental restrictions and modeler are warranted: A person who has memorized how to walk to the bathroom in the dark relying only on feel uses a memorized model of the environment to reach his or her goal. A mobile robot system is much like this person in that the limited sensing capabilities are represented as sensor processing deficiencies.

A *direct effector* (commonly referred to simply as an effector) in a mobile robot can be a movable appendage (manipulator) or a locomotive device (wheels). Certain sensors can be considered *active* since they emanate waves or particles into the environment as part of the sensing process. An example of such a sensor is the ultrasonic range finder. The ultrasonic range finder sends a high frequency impulse into the surrounding environment to stimulate a reflection and eventual detection of the reflected wave.

An *indirect effector* does not change the platform's physical configuration but does affect the environment. For example, an infra-red elevator interface would include a transmitter on the platform and a detector on the elevator control panel. The robot could request elevator service without making contact with the up/down buttons. If a building containing a robot was "wired" to the robot's supervisor computer, then the supervisor could pass a request to the elevator system directly when the robot and elevator come into close proximity. The effector which would have been used to control the elevator is no longer needed. The presence of the robot near the elevator and the intent to ride on the elevator will have caused a *virtual effector event* in the latter variation. There are currently no such events implemented in the LLAMA and G2 system since it is robo-centric.

As robots become more complex the need for self-diagnosis arises. There is no fundamental difference between the robot and its environment to the robot hardware/software and its (robot) environment. A robot is composed of different hardware and software modules that attempt to function as a seamless whole. If a module fails or falters, it is imperative that the supervisor is made aware of the fact. Each module may contain sensors that determine the condition of the module. Such sensors are no different from the sensors used to sense the robot's environment. We can refer to these as *system sensors*.

These sensors can be further sub-classed, but current robot technology does not include these variations so the one term will suffice. Similarly, *system effectors* are plausible devices, but are rarely found in a robot. A consistent definition for a system effector is *any device which causes a change in another device using a means other than a physically guided interface*. An I/O port or optical fiber interface would not qualify under this definition. If these were included in the definition, virtually any software to hardware interface would qualify as an effector! Changes to system parameters or behavior are part of a class called *system interface control*.

Future mobile robots may contain modules, perhaps miniature repair robots, that monitor and adjust system performance. LLAMA can be configured to produce either solicited or unsolicited system sensor reports. Additionally, a rudimentary "self-repair" can be implemented: if some event causes an emergency stop condition, such as running into an object, a macro can be designed which will determine the stopped condition and reset the emergency-stop automatically.

Commands and Reports

A mobile robot, regardless of any constraints placed on its operation, must have the capabilities for movement and sensing. According to our definition, a robot controlled indirectly by a human would probably report sensor data; however, if the platform contained no sensors, the person controlling the robot would have to substitute his or her sight for this purpose. Another sensor technique that relies on neither the platform sensors nor human sight is known by the term, "God's Eye." Modern airborne fighter craft display a top-down view of the combat theater for the pilot.

Usually this visual information is provided by support radar aircraft and received and displayed in the fighter. This example illustrates an important issue in robotics: Should a robotics system rely only on the platform's sensors or all available sensors in the robot's environment? A robot that has access to global sensors is termed *enviro- centric*. A properly designed robo-centric robot has the advantage of being completely self-contained and autonomous. If a general mobile robot is required, it could be designed in an enviro-centric manner. A robo-centric system would be autonomous but could enhance its knowledge of the world by obtaining enviro-centric data upon request. Note that the robo-centric system does not rely on enviro-centric data and can function well without it.

Since a robo-centric system can only rely on its own sensors, in an open autonomous system sensor data must return from the implementor to the supervisor. This is the basis for reports. Similarly, supervisor to implementor control is achieved through the use of commands. One advantageous side effect of a robo-centric open autonomous system is that the platform can be modeled as a black box. The platform has one input and one output to the stationary system, which lends itself to such an interpretation. A greater difficulty is how to model the environment. Either the environment must be modeled as part of the black box, or the environmental model must somehow interface to the model of the robot's sensors and actuators.

The Mobile Robot System Classes

Most robot hardware can be controlled through software. The hardware classes of effectors and sensors are usually controlled through some software function, either a macro or primitive in the LLAMA system, and executed when the corresponding command arrives from G2. Most of these software functions can be grouped into like classes. A hierarchical class description is a reasonably clear method of describing the command and report structure of LLAMA; however, dependencies occur across classes and certain of these relationships might be obfuscated. For example, all drive motor commands can be grouped together but are not since each is grouped under appropriate effector, sensor and system headings (classes). An additional class description, called the *functional* class, supports the programming and operational features of LLAMA. The complete class hierarchy. This class description. The relationship between hardware control through software functions.

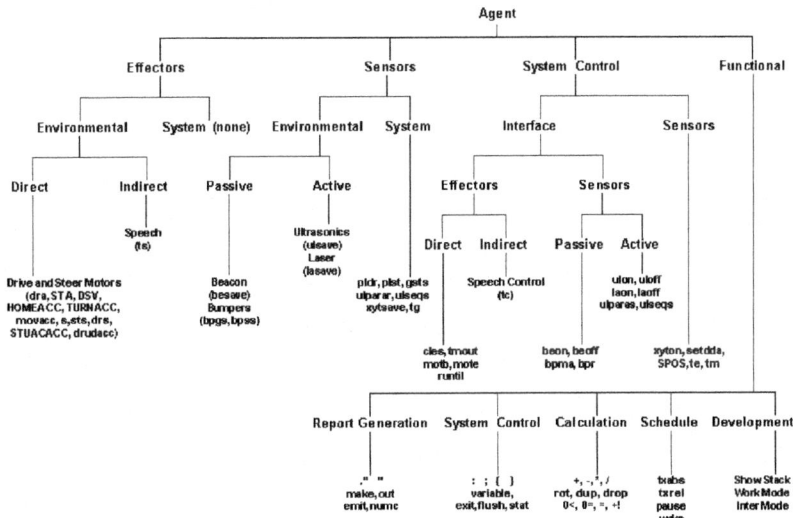

Fig. : Agent Classes. The mobile robot agent classes are presented in a heirarchical format. Due to space limitations, the Functional class has been placed near the bottom of the diagram.

STEREO VISION FOR MOBILE ROBOTICS

Two eyes or cameras looking at the same scene from different perspectives provide a mean for determining three-dimensional shape and position. Scientific investigation of this effect (called variously stereo vision, stereopsis or single vision) has a rich history in psychology, biology and more recently, in the computational model of perception. Stereo is an important method for machine perception because it leads to direct depth measurements. Additionally, unlike monocular techniques, stereo does not infer depth from weak or unverifiable photometric and statistical assumptions, nor does it require specific detailed objects models. Once stereo images have been brought into point-to-point correspondence, recovering depth by triangulation is straightforward.

Current range imagers have achieved real-time or near real-time performance on images of modest size. For example, stereo algorithms on standard hardware are capable of returning dense 128 x 128 range images at 10 Hz, while scanning laser range-finders can operate at 2 Hz on 256 x 256 images. To take advantage of these devices, researchers have proposed numerous methods for extracting 3-D information from range images. These methods operate either in 3-D Cartesian space (*volumetric* representations) or in a 2.5-D range image space (*contour map method*). Contour map methods are particularly attractive for computation bound applications such as mobile robots.

We begin with a discussion of the geometric basis and a computational model for stereo vision. Next, we briefly describe biological aspects of depth perception and a contour map method for depth processing. Finally, we present an obstacle avoidance technique for mobile robots using real-time stereo vision.

Stereo Vision

The geometric basis key problem in stereo vision is to find *corresponding* points in stereo images. Corresponding points are the projections of a single 3D point in the different image spaces. The difference in the position of corresponding points in their respective images is called *disparity*. Disparity is a function of both the position of the 3D scene point and of the position, orientation, and physical characteristics of the stereo devices (*e.g.* cameras).

In addition to providing the function that maps pair of corresponding images points onto scene points, a camera model can be used to constraint the search for corresponding image point to one dimension. Any point in the 3D world space together with the centers of projection of two cameras systems, defines an *epipolar plane*. The intersection of such a plane with an image plane is called an *epipolar line*. Every point of a given epipolar line must correspond to a single point on the corresponding epipolar line. The search for a match of a point in the first image may therefore be reduced to a one-dimensional neighborhood in the second image plane (as opposed to a 2D neighborhood).

When the stereo cameras are oriented such that there is a known horizontal displacement between them, disparity can only occur in the horizontal direction and the stereo images are said to be *in correspondence*. When a stereo pair is in correspondence, the epipolar lines are coincident with the horizontal scan lines of the digitized pictures.

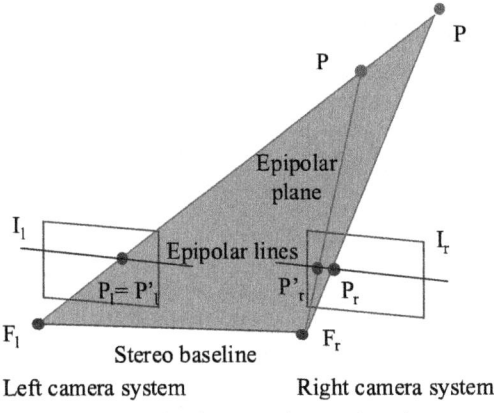

Fig. : Epipolar lines and epipolar planes.

Ideally, one would like to find the correspondence of every individual pixel in both images of a stereo pair. However, it is obvious that the information content in the intensity value of a single pixel is too low for unambiguous matching. In practice, continuous areas of image intensity are the basic units that are matched. This approach (called *area matching*) usually involves some form of *cross-correlation* to establish correspondences.

Matching

The main problem in matching is to find an effective definition of what we call a *valid correlation*.

Correlation scores are computed by comparing a fixed window in the first image against a shifting window in the second. The second window is moved in the second image by integer increments along the corresponding epipolar line and a correlation score curve is generated for integer disparity values. The measured disparity can then be taken to be the one that provides the largest peak.

To quantify the similarity between two correlation windows, we must choose among many different criteria that produce reliable results in a minimum computation time. We denote by $I_1(x,y)$ and $I_2(x,y)$ the intensity value at pixel (x,y). The correlation window has dimensions

$(2n+1)\times(2m+1)$. Therefore, the indexes which appear in the formula below vary between -n and +n for the i-index and between -m and +m for the j-index :

$$C_1(x,y,\Delta) = \frac{\sum_{i,j}[I_1(x+i,y+j) - I_2(x+\Delta+i,y+j)]^2}{\sqrt{\sum_{i,j}I_1^2(x+i,y+j)} \times \sqrt{\sum_{i,j}I_2^2(x+\Delta+i,y+j)}}$$

It is important to know if a match is reliable or not. The form of the correlation curve (for example C_1) can be used to decide if the probability of the match to be an error is high or not. Indeed, errors occur when a wrong peak slightly higher than the right one is chosen. Thus, if in the correlation curve we find several peaks with approximately the same height, the risk of choosing the wrong one increases, especially if the image is noisy. However, a *confidence coefficient* σ, proportional to the difference of height between the most important peaks may be defined. Other important information may also be extracted from the correlation curve as, for instance, bland areas.

Human Depth Perception

For human beings, correlation is only a *local* mechanism of stereoscopic vision. However, imagine the following experiment:

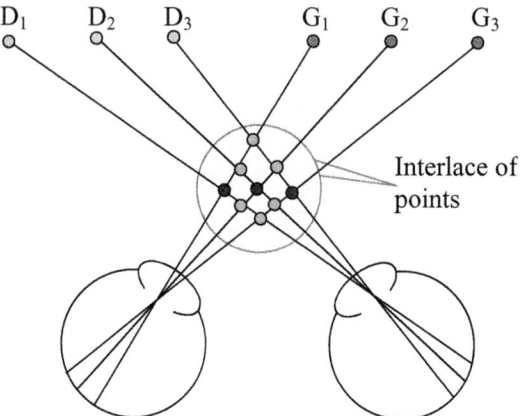

Fig. : Stereoscopic fusion false target problem.

A stereoscopic system displays the set of points D_1, D_2, D_3 for the right eye and G_1, G_2, G_3 for the left one. An observer should be able to see any interlace of points (grey points in the light grey area) but, instead, they all succeed to the dark ones. This experiment shows that a *global* mechanism, based on criteria other than local correlation, is used. Among these ones, the following are taken in account:

- A principle of correlation based on the contours of the image;
- A mechanism of cognitive interpretation which has, in some cases, more priority than the local mechanism of stereo vision;

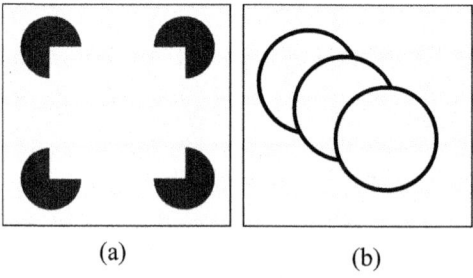

(a)   (b)

Fig. : (a) "Illusory" contours defining a square giving the impression that this shape is placed in front of 4 circles, (b) interposition principle (cognitive interpretation).

- A mechanism of pictorial clues of depth (relative size, relative height, perspective, shade, "fog" effect and interposition;
- A principle of dynamic clues, such as motion parallax;
- Other mechanisms such as correlation of frequency filtered images;

This list is not exhaustive but presents the most significant criteria belonging to the global system of the human stereoscopic vision.

In summary, the human stereo system uses a number of interesting methods, which work together to recover depth. On one hand, this system is very powerful because even using one eye, it is possible to perceive depth. On the other hand, it is also very *subjective* because a trompe-l'œil can fool our perceptive system.

Contour Map Method

For many applications, working in 3D Euclidean space turns out to be unnecessary and difficult to manage. To reduce the amount of data which has to be processed, we introduce a method of quantifying volumes that allows us to manipulate range images directly, without having to first transform to 3-D space. The method is similar to the use of contour maps to represent elevations; hence, we call it the *contour method*.

A contour represents the elevation at a particular height; all terrain between one contour line and the next is at an elevation between that represented by the contour lines. Contour creation can be visualized as a set of planes parallel to the ground at specified heights, intersecting the terrain. These *cutting planes* induce a quantization of the 3-D space based on elevation. Our approach uses this basic idea, and consists of the following steps:

1. Constructing a set of volumes in 3-D space using a set of cutting surfaces (not necessarily planar);
2. Projecting the cutting surfaces back to the range image to induce a quantization of the range data;
3. Using the quantized range image to construct terrain models or other abstractions;

In cases where the desired segmentation is relative to the sensor viewpoint, the first two steps can be achieved off-line, leading to significant computational savings, especially when the cutting surfaces are complicated. In addition, step 3 can often be performed in the range image space, which is much more efficient than working in the volumetric space. Also, in contrast to the *grid-based approaches*, the cutting surfaces need not be regular, and can be sized to take in account the precision and error characteristics of the range data.

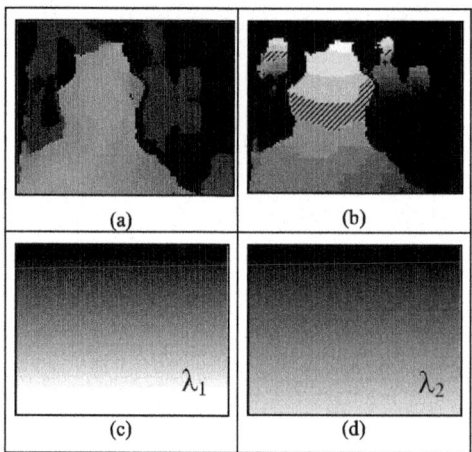

Fig. : (a) Range image where light (green) pixels represent points closer than dark ones, (b) Elevation map composed of the superposing of 8 contours (a single contour is represented with a special pattern), (c) and (d) projection of the cutting surfaces back in the range image for the special contour of image b.

Mobile Robot Obstacle Avoidance

The contour method is well suited for use with the *vector field histogram* (VFH) algorithm for mobile robot obstacle avoidance. Originally developed with sonar sensors, the method used three steps:

1. A regular 2-D *histogram grid* in plan view, holding the results of sonar sensor readings around the robot. The value of each grid point represents the number of sonar readings that indicated an object within the point;

2. A *polar histogram* is computed from the histogram grid, with k regular angular sectors instead of a rectilinear grid. The value h_k of each sector in the polar histogram represents the obstacle density in that direction;

3. Steering and velocity values are extracted from the polar histogram;

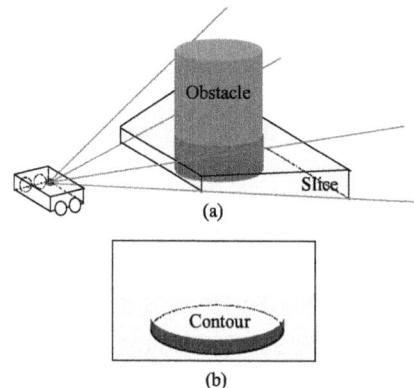

Fig. : Representation of an obstacle (a) in the Cartesian space, (b) in the contour image space.

In range images, each column of the image represents a polar sector whose angular width is determined by the camera parameters. We let the k sectors correspond to the columns of the range image. Thus, we can construct the polar histogram directly from the contour representation, without having to convert to Cartesian space.

The key step is calculating the histogram value h_k for each sector. Roughly speaking, this value represents the probability of finding an obstacle close to the robot in the direction of sector k. The simplest idea is to use a single cutting surface at elevation over the ground plane sufficient to constitute an obstacle for the robot. Any points in the resulting contour are obstacles, and we can use the number of such points in a column and their distance to determine a histogram value.

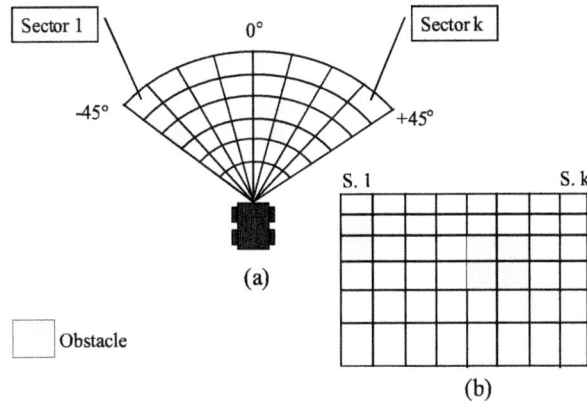

Fig. : Two obstacles (a) in the polar grid, (b) in the contour image grid.

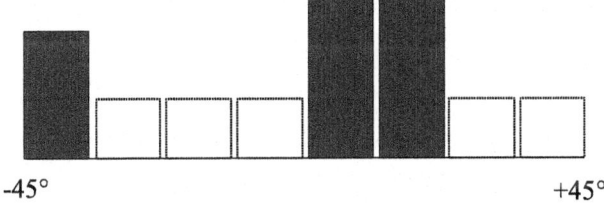

Fig. : Polar histogram corresponding the obstacles.

The details of the weighting scheme we use are not critical; we expect almost any reasonable method that combines distance and number of points will work reasonably well. In our implementation we used a stereo system with disparity as the range metric, and let each contour cell m contributes to its sector value. This measure compensates for the fact that the disparity increases hyperbolically as an object gets closer.

The VFH method was implemented using a small stereo system for range images and a PC for processing the VFH algorithm. The stereo system returned images at a 5 Hz rate, and the VFH processing took less than 10 ms per image to format the polar histogram and extract the desired direction and speed of travel. Data were then sent to a robot navigation program in order to steer a Koala or Pioneer robot.

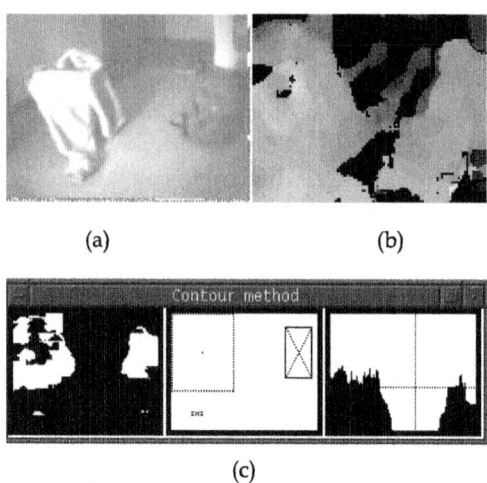

(a)   (b)

(c)

Fig. : Experimental results. (a) image of the scene, (b) corresponding disparity image, (c) from left to right: contour, object detection and polar histogram (the vertical line corresponds to the direction followed by the robot and the horizontal line, its speed).

A calculation of the polar histogram from a typical range image and a single-contour segmentation. In this case, the sensor covers about a 70 degree angle, and each sector is about 0.5 degrees. Image (a) is an intensity image of the scene, and (b) shows the disparity map computed by a stereo system. Brighter green values are higher disparities, hence closer to the camera. The final set of images (c) shows the contour (left side), a segmentation of some obstacles (middle), and finally the polar histogram. The middle of the histogram is straight along the camera optical axis, and the vertical line indicates the direction of travel that the VFH algorithm has found. From the picture, this is the direction through the open door.

Because of the small sector size, there is considerable variation in adjacent histogram values, and the result could benefit from low-pass filtering.

Additional enhancements in the construction of the polar histogram from contours are under study, among them:

Ground plane detection. A contour representing the ground plane would give an indication that there actually was a reasonable path in front of the robot.

Holes. A contour underneath the ground plane could be used to check for holes near the robot. The addition to the histogram would be the same as for positive-elevation obstacles; Small-height obstacles. Instead of a single contour at an appropriate height for obstacles, several could be positioned starting from just over the ground plane. The contribution of elements from the lower contours would be weighted by a fraction q depending on their height. Thus, the robot would prefer smooth terrain to bumpy, even though it could negotiate the latter.

CONTROLLING A MOBILE ROBOT WITH VISUAL PROLOG

The practical realization of a simple declarative model for controlling a technical object – a mobile robot, which was built at the Robotics and Artificial Intelligence Laboratory of the State Polytechnic Museum under Technics PhD Valery Karpov and Technics PhD Dmitry Dobrynin. Students and schoolchildren develop and program their own robots there. As an AI student I participate in this laboratory and would like to present my first robot.

Architecture of the System

The model of the robot's behavior is similar to the behavior of a living creature.

The system can be divided into the low-level actions, which are processed by the robot himself, and the high-level actions, which are suggested by a Prolog program on a stationary computer.

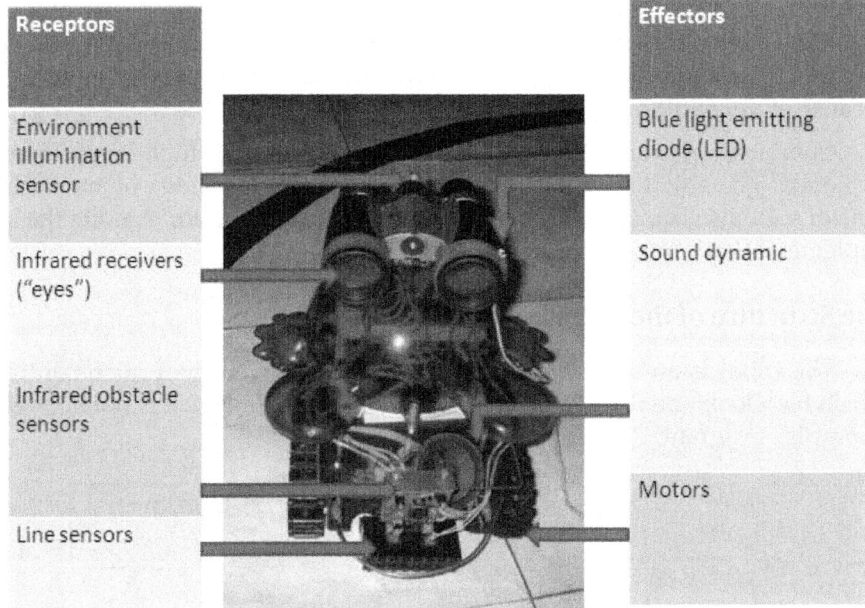

Fig. : Robot's receptors and effectors.

Low-level actions include: unconditional reflexes (evading an obstacle, getting frightened by a sudden light, following the path on the floor, following an IR emitter, defining whether he is at home ("home" is a grey spot on the floor)), simple actions (lighting a LED, crying), decomposition of high-level commands to low-level ones. All low-level actions are programmed into the robot's microcomputer on C language.

High-level commands are planning, guiding, database maintaining and the world map building.

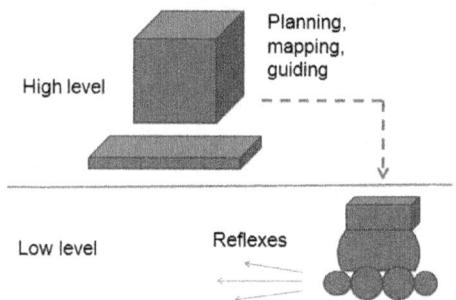

Fig. : Two levels of robot's control.

If robot doesn't receive any message from the computer during several seconds, he considers himself "lost" and wanders randomly, only reacting to the environmental exposure with his reflexes. When the computer takes control, robot becomes passive and obeys the computer.

The advantage of such system is that even in the case of a mistake or a controversial command from computer, robot can still evade dangerous situations because of unconditional reflexes.

Combination of simple low-level reflexes and complex high-level command sequences gives us the model of a living organism. The idea of hierarchy in control systems is quite old, but not widely spread in robotics. Among the latest implementations, must be noted.

The Structure of the Robot

The robot is an autonomous mobile device, driven by a microcontroller, which has clock rate 7 MHz and flash memory of 8 Kb, which is not enough for a complex program.

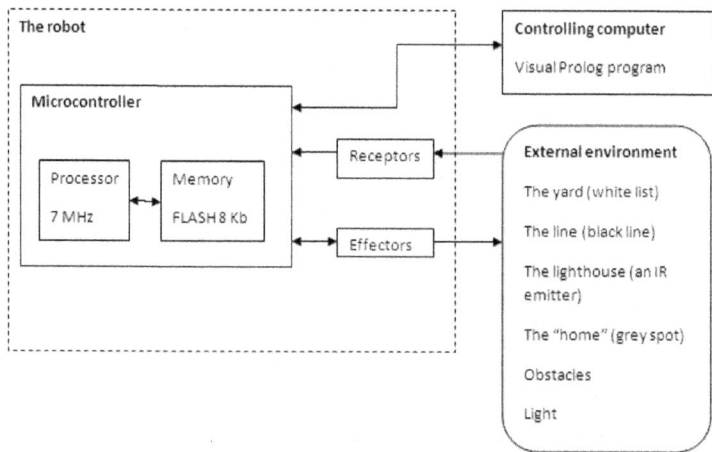

Fig. : The model of the system.

Robot has different sensors (receptors): environment illumination sensor, infrared receivers ("eyes"), infrared obstacles sensors, line photo sensors; effectors, which influence the environment: motors, light-emitting diode, sound dynamic.

Robot sends information about his movements, obtained by encoders on both driving wheels, so we can know the length of movement vector and its angle relatively to the robot's starting position. Changing polar coordinates to Cartesian, computer tries to draw robot's position on the map and assert the coordinates of all significant objects to the database to use them later for navigation.

Communication Program

To establish a connection with robot, a terminal program on Visual Prolog is required. Ben Hooijenga wrote such program under VIP 7.1 CE, it's available on PDC forum and on Visual Prolog Wiki. The program is easy to apply and has a convenient GUI for setting all the parameters. The program makes it possible to send and receive bytes *via* RS-232 interface and process communicstion events. A declarative system for the robot was added to this program. To make a communications channel between robot and computer, a protocol was developed.

Description of the Protocol

To avoid synchronization problems, a firm scheme involving timer was specified. Each 100 milliseconds a communication act takes place. Computer sends a command of sensors polling, encoded with one byte. Robot receives it and answers with two bytes, first of which is the robot's state and information about his movements. Second byte represents signals on robot's sensors. Computer decides what robot should do next, and answers with a command, typically, of one byte.

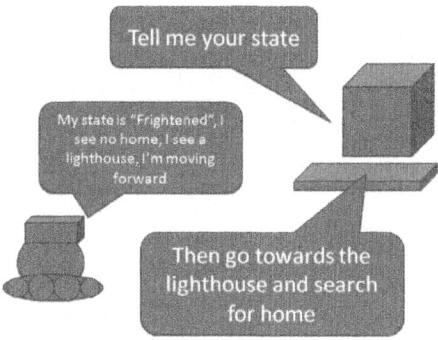

Fig. : Example of communication.

An example of interaction between robot and computer

After being turned on, robot is moving around for some time to adjust his sensors. In the initial state robot moves randomly, until the computer tells him to look for the line. Having found the line, robot starts to follow it.

If a bright flash of light frightens robot, he "cries" and "panics", moving around randomly, till the computer orders him to turn LED on and follow towards the lighthouse (the lighthouse stands in the middle of "home").

Robot looks for the lighthouse and goes towards it, checking if he is already at "home". If robot reaches home, a command from computer tells him to return to the line.

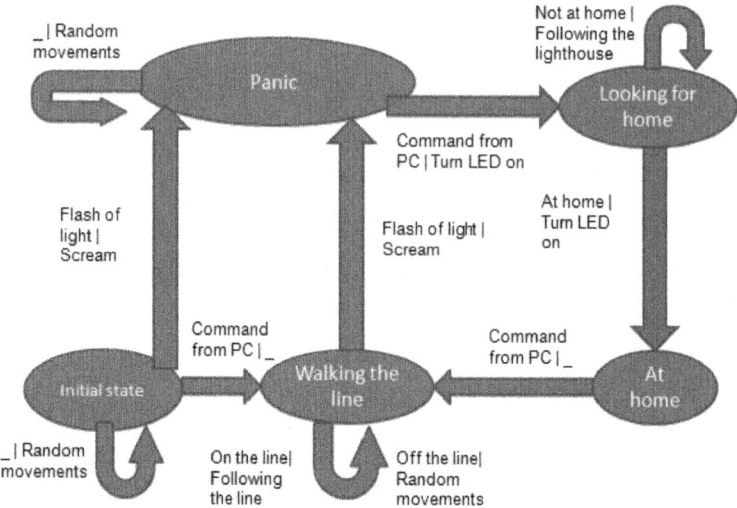

Fig. : The model of robot's behavior. Red transitions can only be made after a computer's command and they do not contain in robot's memory.

Plans

Now we are working on next projects. Among our projects are continuation of work on museum robot-excursion guide and building more complicated robots. My second robot will be a robot-navigator.

Chapter 5

ROBOT PROGRAMMING AND LANGUAGES

ROBOT SOFTWARE

Robot software is the set of coded commands that tell a mechanical device and electronic system, known together as a robot, what tasks to perform. Robot software is used to perform autonomous tasks. Many software systems and frameworks have been proposed to make programming robots easier.

Some robot software aims at developing intelligent mechanical devices. Common tasks include feedback loops, data filtering, control, pathfinding, and locating.

Robot software is the coded commands that tell a mechanical device (known as a robot) what tasks to perform and control its actions. Robot software is used to perform tasks and automate tasks to be performed. Programming robots is a non-trivial task. Many software systems and frameworks have been proposed to make programming robots easier.

Some robot software aim at developing intelligent mechanical devices. Though common in science fiction stories, such programs are yet to become common-place in reality and much development is yet required in the field of artificial intelligence before they even begin to approach the science fiction possibilities. Pre-programmed hardware may include feedback loops such that it can interact with its environment, but does not display actual intelligence.

Data flow programming techniques are used by most robot manufacturers, and is based on the concept that when the value of a variable changes, the values of other variables affected should also change. A programming language that incorporates data flow principles is called a data flow language. In addition to numeric processing, data flow languages also incorporate functional concepts. Unlike other programming languages which use imperative programming, data flow programming is modeled as a sequence of functions.

With any programming software, the state of a program at any given time is an important consideration. The state provides an indication of the various conditions at a particular instant. In order to function properly, most programming languages require a significant amount of state information. This information is invisible to the programmer.

Another key concept – which is associated with any type of robot programming, is the concept of run-time. When a program is running, or executing, it is said to be in run-time. The term run-time is also used as a short form when referring to a run-time library, which is a library of code instructions used by a computer language to manage a program written in the language. The term is also used by software developers to specify when errors in a program can occur. A runtime error is an error that happens while the program is executing. For example, if a robot arm was programmed to turn left, and it turned right, then that would be a runtime error.

The software architecture of a system consists of the various software components used to design and operate the software. All programming methods rely on software architecture as a method of organizing a software system since it not only provides communication support but is also a critical component in hardware and software interfaces.

Examples of Programming Languages for Industrial Robots

Due to the highly proprietary nature of robot software, most manufacturers of robot hardware also provide their own software. While this is not unusual in other automated control systems, the lack of standardization of programming methods for robots does pose certain challenges. For example, there are over 30 different manufacturers of industrial robots, so there are also 30 different robot programming languages required. Fortunately, there are enough similarities between the different robots that it is possible to gain a broad-based understanding of robot programming without having to learn each manufacturer's proprietary language.

Some examples of published robot programming languages are shown below.

Task in plain English:

```
Move to P1 (a general safe position)
Move to P2 (an approach to P3)
Move to P3 (a position to pick the object)
Close gripper
Move to P4 (an approach to P5)
Move to P5 (a position to place the object)
Open gripper
Move to P1 and finish
```

VAL was one of the first robot 'languages' and was used in Unimate robots. Variants of VAL have been used by other manufacturers including Adept Technology. Stäubli currently use VAL3.

Example program:

```
PROGRAM PICKPLACE
1. MOVE P1
2. MOVE P2
3. MOVE P3
4. CLOSEI 0.00
5. MOVE P4
6. MOVE P5
7. OPENI 0.00
8. MOVE P1
.END
```

Epson RC+ (example for a vacuum pickup)

```
Function PickPlace
   Jump P1
   Jump P2
   Jump P3
   On vacuum
   Wait.1
   Jump P4
   Jump P5
   Off vacuum
   Wait.1
   Jump P1
Fend
```

ROBOFORTH (a language based on FORTH).

```
: PICKPLACE
P1
P3 GRIP WITHDRAW
P5 UNGRIP WITHDRAW
P1
;
```

(With Roboforth you can specify approach positions for places so you do not need P2 and P4.)

Clearly the robot should not continue the next move until the gripper is completely closed. Confirmation or allowed time is implicit in the above examples of CLOSEI and GRIP whereas the On vacuum command requires a time delay to ensure satisfactory suction.

OTHER ROBOT PROGRAMMING LANGUAGES

Visual Programming Language

The software system for the Lego Mindstorms EV3 robots is worthy of mention. It is a graphical user interface (GUI) written with LabVIEW. The approach is to start with the program rather than the data. The program is constructed by dragging icons into the program area and adding or inserting into the sequence. For each icon you then specify the parameters (data). For example for the motor drive icon you specify which motors and by how much they move. When the program is written it is downloaded into the Lego NXT 'brick' (microcontroller) for test.

Scripting Languages

A scripting language is a high-level programming language that is used to control the software application, and is interpreted in real-time, or "translated on the fly", instead of being compiled in advance. A scripting language may be a general-purpose programming language or it may be limited to specific functions used to augment the running of an application or system program. Some scripting languages, such asRoboLogix, have data objects residing in registers, and the program flow represents the list of instructions, or instruction set, that is used to program the robot.

Table : Programming languages in industrial robotics.

Robot brand	Language name
ABB	RAPID
Comau	PDL2
Fanuc	Karel
Kawasaki	AS
Kuka	KRL
Yaskawa	Inform

Programming languages are generally designed for building data structures and algorithms from scratch, while scripting languages are intended more for connecting, or "gluing", components and instructions together. Consequently, the scripting language instruction set is usually a streamlined list of program commands that are used to simplify the programming process and provide rapid application development.

Parallel Languages

Another interesting approach is worthy of mention. All robotic applications need parallelism and event-based programming. Parallelism is where the robot does two or more things at the same time. This requires appropriate hardware and software. Most programming languages rely on threads or complex abstraction classes to handle parallelism and the complexity that comes with it, like concurrent access to shared resources. URBI provides a higher level of abstraction by integrating parallelism and events in the core of the language semantics.

```
whenever(face.visible)
{
   headPan.val += camera.xfov * face.x
   &
   headTilt.val += camera.yfov * face.y
}
```

The above code will move the headPan and headTilt motors in parallel to make the robot head follow the human face visible on the video taken by its camera whenever a face is seen by the robot.

Robot Application Software

Regardless which language is used, the end result of robot software is to create robotic applications that help or entertain people. Applications include command-and-control and tasking software. Command-and-control software includes robot control GUIs for tele-operated robots, point-n-click command software for autonomous robots, and scheduling software for mobile robots in factories. Tasking software includes simple drag-n-drop interfaces for setting up delivery routes, security patrols and visitor tours; it also includes custom programs written to deploy specific applications. General purpose robot application software is deployed on widely distributed robotic platforms.

Safety Considerations

Programming errors represent a serious safety consideration, particularly in large industrial robots. The power and size of industrial robots mean they are capable of inflicting severe injury if programmed incorrectly or used in an unsafe manner. Due to the mass and high-speeds of industrial robots, it is always unsafe for a human to remain in the work area of the robot during automatic operation. The system can begin motion at unexpected times and a human will be unable to react quickly enough in many situations, even if prepared to do so. Thus, even if the software is free of programming errors, great care must be taken to make an industrial robot safe for human workers or human interaction, such as loading or unloading parts, clearing a part jam, or performing maintenance.

SIMULATION OF ROBOT DYNAMICS AND CONTROL SOFTWARE

Most automated manufacturing tasks are done by special-purpose machines that are designed to perform prespecified functions in a manufacturing process. The inflexibility of these machines makes the computer-controlled manipulators more attractive and cost-effective in various manufacturing and assembly tasks. Today's industrial robots, though controlled by mini-/microcomputer, are basically simple positional machines. A given task had been exacted by playing back prerecorded or preprogrammed sequences of motions that have been previously guided or taught by a user with a hand-held control/teach box. Moreover, because the robots are equipped with few or no external sensors (both contact and non-contact), they cannot obtain vital information about their working environment. More research needs to be directed towards improving the overall performance of the manipulator systems, and one way is through the study of robot arm kinematics, dynamics, and control.

The purpose of robot arm control is to maintain the dynamic response of a computer-based manipulator in accordance with some prespecified system performance and goals. In general, the control problem consists of obtaining suitable dynamic models of the physical robot arm for designing the controller and specifying corresponding control laws or strategies to achieve the desired system response and performance.

Simulink is a software package for modeling, simulating, and analyzing dynamic systems. It supports linear and nonlinear systems, modeled in continuous time, sampled time, or a hybrid of the two. Systems can also be multirate, *i.e.*, have different parts that are sampled or updated at different rates. The MatLab and Simulink environments are integrated into one entity, and thus analyzing, simulation, and revising the models can be done in either environment at any point, and if Behind every robotic movement is a series of complex geometric evaluations and equations that describe motion dynamics. Conformance to system goals must also be considered; control prototyping with Simulink allows a designer to conceptualize control solutions by means of block diagram driven environment.

There is no need to use the production software and hardware during the design process thereby cutting down the cost in design. moreover, the designer can also refine the system model iteratively and tune controller parameters. these features help to reduce the implementation time, which is important in many industry situation. teaching of rapid control prototyping concept is getting acceptance by many academics in leading institutions.

In this work the simulation of dynamic and control of robot manipulator was done by using MatLab/Simulink software; were the robotics toolbox for MatLab was used in the modeling and simulation process. Two types of control problems was studied by using Simulink namely: feedforward control and computed torque control to check the advantage of the of using Simulink in robot modeling and simulating the robot model chosen as industrial robot.

Robot Dynamics

Robot arm dynamics deals with the mathematical formulations of the equations of robot arm motion. The dynamic equations of manipulator motion are a set of equations describing the dynamic behavior of the manipulator. Such equations of motion are useful for computer simulation of robot arm motion, the design of suitable control equations for a robot arm, and the evaluation of the kinematic design and structure of a robot arm.

Various approaches are available to formulate robot arm dynamics, such as the Lagrange-Euler, the Newton-Euler, the recursive Lagrange-Euler, and the generalized d'Alembert principle formulations. Deriving the dynamic model of a manipulator using the L-E method is simple and systematic. The resultant equations of motion, excluding the dynamics of the electronic control device and the gear friction, are a set of second order, coupled nonlinear differential equations.

One approach that has the advantage of both speed and accuracy is based on the N-E vector formulation was used in this work. The derivation is simple, although messy, and involves vector cross-product terms. The resultant dynamic equations, excluding the dynamics of the control device and the gear friction, are a set of forward and backward recursive equations. These equations can be applied to the robot links sequentially.

There are two problems related to manipulator dynamics that are important to solve:

- inverse dynamics in which the manipulator's equations of motion are solved for given motion to determine the generalized forces. and
- direct dynamics in which the equations of motion are integrated to determine the generalized coordinate response to applied generalized forces.

The equations of motion for an n-axis manipulator are given by:

$$\tau = M(q)\ddot{q} + C(q,\dot{q})\dot{q} + G(q)$$

where: if we have 4DOF $q = \begin{bmatrix} \theta_1 & \theta_2 & \theta_3 & \theta_4 \end{bmatrix}^T$, $\dot{q} = \dfrac{dq}{dt}$, $\ddot{q} = \dfrac{d^2q}{dt^2}$

Obtaining the dynamic equations of motion using the MatLab program(robotics toolbox) was very powerful process; the toolbox use N-E approach to compute the equation of motion by feeding the program by necessary data about the robot system by using the functions(dyn.m, robot.m) to introduce the robot object in the MatLab program, the robot object then can be used in the MatLab program to define Simulink blocks. Because of the nature of the formulation and the method of systematically computing the torques, computations are much simpler, allowing a short computing time. With this algorithm, about three milliseconds are needed to compute the feedback joint torques per trajectory set point.

Robot Control

Robot control is the spine of robotics. It consists in studying how to make a robot manipulator do what it is desired to do automatically; hence, it includes in designing robot controllers. Typically, these take the form of an equation or an algorithm which is realized *via* specialized computer programs. Then, controllers form part of the so-called robot control system which is physically constituted of a computer, a data acquisition unit, actuators (typically electrical motors), the robot itself and some extra "electronics". In this work two types of control problems was studied feedforward control and computed torque control.

Computed Torque Control

In order to overcome drawbacks of the PD controller, a more sophisticated scheme in which the magnitude of the nonlinear disturbing and loading torques is computed using the dynamic equations and used to compensate these disturbances by means of a feedforward may be employed. It must be noted that the basic control method is still PD controller with both position and velocity feedback.

The dynamic model (1) that characterizes the behavior of robot manipulators is in general, composed of nonlinear functions of the state variables (joint positions and velocities). This feature of the dynamic model might lead us to believe that given any controller, the differential equation that models the control system in closed loop should also be composed of nonlinear functions of the corresponding state variables. Nevertheless, there exists a controller which is also nonlinear in the state variables but which leads to a closed-loop control system which is described by a linear differential equation. This controller is capable of fulfilling the motion control objective.

The computed-torque control law is given by:

$$\tau = M(q)[\ddot{q}_d + k_v \dot{\tilde{q}} + k_p \tilde{q}] + C(q,\dot{q}) + G(q)$$

where the gains k_p, k_v are chosen to meet some specific properties of the system and these gains should be adjusted to reduce the errors. The block diagram that corresponds to computed-torque control of robot manipulators.

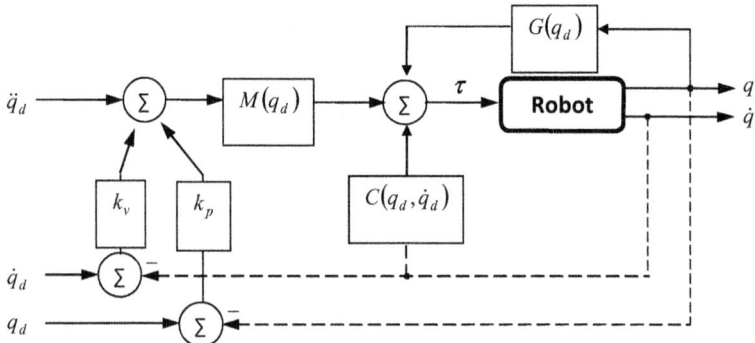

Fig. : Block-diagram: computed torque control.

Feedforward Control

Among the conceptually simplest control strategies that may be used to control a dynamic system we find the so-called open-loop control, where the controller is simply the inverse dynamics model of the system evaluated along the desired reference trajectories. For the case of linear dynamic systems, this control technique may be roughly presented as follows: by using (1) applying a torque τ at the input of the robot, the behavior of its outputs q and $\dot{q}$ are governed by:

$$\frac{d}{dt}\begin{bmatrix} q \\ \dot{q} \end{bmatrix} = \begin{bmatrix} \dot{q} \\ M(q)^{-1}[\tau - C(q,\dot{q}) - G(q)] \end{bmatrix}$$

If the behavior of the outputs q and $\dot{q}$ need to be equal to that specified by q_d and $\dot{q}_d$ respectively, it seems reasonable to replace q, $\dot{q}$ and $\ddot{q}$ by $q_d, \dot{q}_d$, and $\ddot{q}_d$ in the Eq. and to solve for τ. This reasoning leads to the equation of the feedforward controller, it can be expressed as:

$$\tau = M(q_d)\ddot{q}_d + C(q_d,\dot{q}_d)\dot{q}_d + G(q_d)$$

The block-diagram corresponding to a robot under feedforward control is presented; the control action τ does not depend on q nor on $\dot{q}$, that is, it is an open loop control.

The wide practical interest in incorporating the smallest number of computations in real time to implement a robot controller has been the main motivation for the PD plus feedforward control law, Feedforward control may be modified by the addition, of a *feedback* Proportional–Derivative (PD) term. given by:

$$\tau = M(q_d)\ddot{q}_d + C(q_d,\dot{q}_d)\dot{q}_d + G(q_d) + k_v\dot{\tilde{q}} + k_p\tilde{q}$$

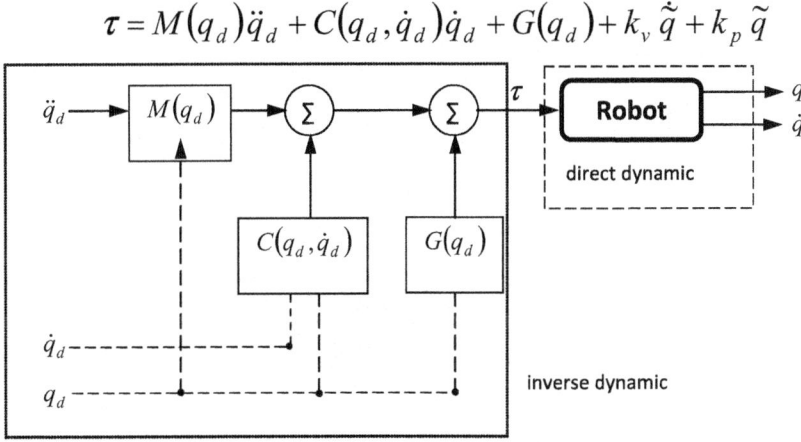

Fig. : Block-diagram: feedforward control.

Simulink Model

The robot model studied in this work is the IRIS robot manipulator where all kinematic and dynamic parameters, the dynamic model for this robot system is a highly nonlinear differential equation due to the coupled between different degrees of freedom(4DOF). The modeling of the robot system was done using Robotics Toolbox in the MatLab program, using the functions(dyn.m, robot.m) to introduce the robot object in the program and then the robot object can be used with two way dynamics; The forward recursion propagates kinematics information- such as angular velocities, angular accelerations, linear accelerations, total forces and moments exerted at the center of mass of each link-from the base reference frame (inertial frame) to the end-effector. The backward recursion propagates the forces and moments exerted on each link from the end effector of the manipulator to the base reference frame. using the inverse and the direct dynamic functions which are very important in the usage of dynamic block of Simulink in the robot control simulation.

PROGRAMMING LEGO ROBOTS

Writing Your First Program

Building a Robot

The robot we will use throughout this tutorial is a simple version of the top-secret robot that is described on page 39-46 of your constructopedia. We will only use the basis chassis. Remove the whole front with the two arms and the touch sensors. Also, connect the motors slightly different such that the wires are connected to the RCX at the outside. This is important for your robot to drive in the correct direction. Your robot should look like this:

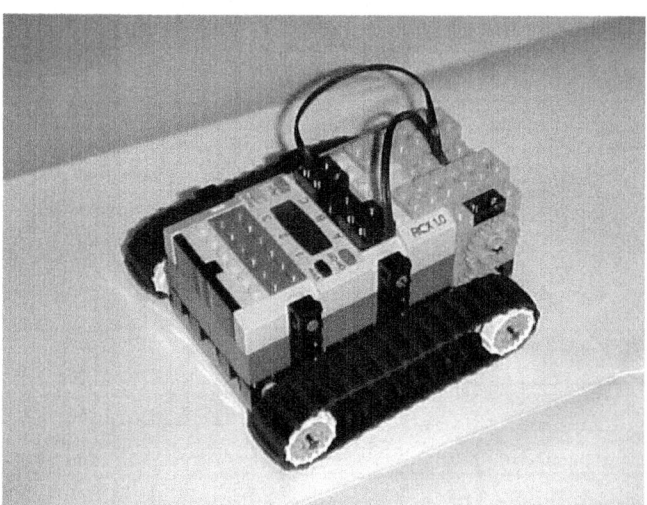

Also make sure that the infra-red port is correctly connected to your computer and that it is set to long range.

Starting Bricx Command Center

We write our programs using Bricx Command Center. Start it by double clicking on the icon BricxCC. (I assume you already installed Bricx Command Center. If not, download it from the web site, and install it in any directory you like.) The program will ask you where to locate the robot. Switch the robot on and press **OK**. The program will (most likely) automatically find the robot. Now the user interface appears as shown below (without a window).

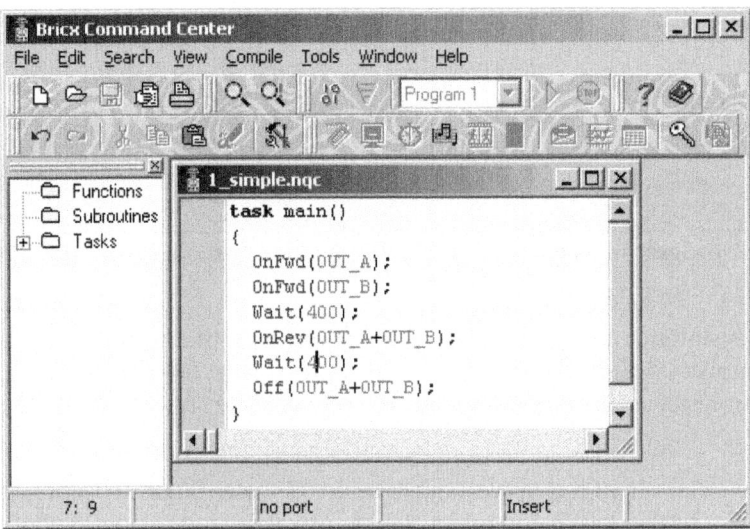

The interface looks like a standard text editor, with the usual menu's, and buttons to open and save files, print files, edit files, *etc.* But there are also some special menus for compiling and downloading programs to the robot and for getting information from the robot. You can ignore these for the moment.

We are going to write a new program. So press the **New File** button to create a new, empty window.

Writing the program

Now type in the following program:

```
task main()
{
  OnFwd(OUT_A);
  OnFwd(OUT_C);
  Wait(400);
  OnRev(OUT_A+OUT_C);
  Wait(400);
  Off(OUT_A+OUT_C);
}
```

It might look a bit complicated at first, so let us analyze it. Programs in NQC consist of tasks. Our program has just one task, named `main`. Each program needs to have a task called `main` which is the one that will be executed by the robot. A task consists of a number of commands, also called statements. There are brackets around the statements such that it is clear that they all belong to this task. Each statement ends with a semicolon. In this way it is clear where a statement ends and where the next statement begins. So a task looks in general as follows:

```
task main()
{
    statement1;
    statement2;
        ...
}
```

Our program has six statements. Let us look at them one at the time:

```
OnFwd(OUT_A);
```

This statement tells the robot to start output A, that is, the motor connected to the output labeled A on the RCX, to move forwards. It will move with maximal speed, unless you first set the speed. We will see later how to do this.

```
OnFwd(OUT_C);
```

Same statement but now we start motor C. After these two statements, both motors are running, and the robot moves forwards.

```
Wait(400);
```

Now it is time to wait for a while. This statement tells us to wait for 4 seconds. The argument, that is, the number between the parentheses, gives the number of "ticks". Each tick is 1/100 of a second. So you can very precisely tell the program how long to wait. So for 4 seconds, the program does do nothing and the robot continues to move forwards.

```
OnRev(OUT_A+OUT_C);
```

The robot has now moved far enough so we tell it to move in reverse direction, that is, backwards. Note that we can set both motors at once using OUT_A+OUT_C as argument. We could also have combined the first two statements this way.

```
Wait(400);
```

Again we wait for 4 seconds.

```
Off(OUT_A+OUT_C);
```

And finally we switch both motors off.

That is the whole program. It moves both motors forwards for 4 seconds, then backwards for 4 seconds, and finally switches them off.

You probably noticed the colors when typing in the program. They appear automatically. The colors and styles used by the editor when it performs syntax highlighting are customizable.

Running the Program

Once you have written a program, it needs to be compiled (that is, changed into code that the robot can understand and execute) and send to the robot using the infra red link (called "downloading" the program). There is a button that does both at once. Press this button and, assuming you made no errors when typing in the program, it will correctly compile and be downloaded.

Now you can run your program. To this end press the green run button on your robot or, more easily, press the run button on your window. Does the robot do what you expected? If not, the wires are probably connected wrong.

Errors in Your Program

When typing in programs there is a reasonable chance that you make some errors. The compiler notices the errors and reports them to you at the bottom of the window, like in the following figure:

```
1_errors.nqc                                          _ □ X
task main()
{
    OnFwd(OUT_D);
    OnFwd(OUT_C);
    Wait(400);
    OnRev(OUT_A+OUT_C);
    Wait(400);
    Of(OUT_A+OUT_C);
}

line 3: Error: undefined variable 'OUT_D'
```

It automatically selects the first error (we mistyped the name of the motor). When there are more errors, you can click on the error messages to go to them. Note that often errors at the beginning of the program cause other errors at other places. So better only correct the first few errors and then compile the program again. Also note that the syntax highlighting helps a lot in avoiding errors. For example, on the last line we typed Of rather than Off. Because this is an unknown command it is not highlighted.

There are also errors that are not found by the compiler. If we had typed OUT_B this would have gone unnoticed because that motor exists (even though we do not use it in the robot). If your robot exhibits unexpected behavior, there is most likely something wrong in your program.

Changing the Speed

As you noticed, the robot moved rather fast. Default the robot moves as fast as it can. To change the speed you can use the command SetPower(). The power is a number between 0 and 7. 7 is the fastest, 0 the slowest (but the robot will still move). Here is a new version of our program in which the robot moves slow:

```
task main()
{
  SetPower(OUT_A+OUT_C,2);
  OnFwd(OUT_A+OUT_C);
  Wait(400);
  OnRev(OUT_A+OUT_C);
  Wait(400);
  Off(OUT_A+OUT_C);
}
```

A More Interesting Program

Our first program was not very spectacular. So let us try to make it more interesting. We will do this in a number of steps, introducing some important features of our programming language NQC.

Making turns

You can make your robot turn by stopping or reversing the direction of one of the two motors. Here is an example. Type it in, download it to your robot and let it run. It should drive a bit and then make a 90-degree right turn.

```
task main()
{
  OnFwd(OUT_A+OUT_C);
  Wait(100);
  OnRev(OUT_C);
  Wait(85);
  Off(OUT_A+OUT_C);
}
```

You might have to try some slightly different numbers than 85 in the second Wait() command to make a precise 90-degree turn. This depends on the type of surface on which the robot runs. Rather than changing this in the program it is easier to use a name for this number. In NQC you can define constant values as shown in the following program.

```
#define MOVE_TIME    100
#define TURN_TIME     85

task main()
{
  OnFwd(OUT_A+OUT_C);
  Wait(MOVE_TIME);
  OnRev(OUT_C);
  Wait(TURN_TIME);
  Off(OUT_A+OUT_C);
}
```

The first two lines define two constants. These can now be used throughout the program. Defining constants is good for two reasons: it makes the program more readable, and it is easier to change the values. Note that Bricx Command Center gives the define statements its own color.

Repeating Commands

Let us now try to write a program that makes the robot drive in a square. Going in a square means: driving forwards, turning 90 degrees, driving forwards again, turning 90 degrees, *etc.* We could repeat the above piece of code four times but this can be done a lot easier with the **repeat** statement.

```
#define MOVE_TIME    100
#define TURN_TIME     85

task main()
{
  repeat(4)
  {
    OnFwd(OUT_A+OUT_C);
    Wait(MOVE_TIME);
    OnRev(OUT_C);
    Wait(TURN_TIME);
  }
  Off(OUT_A+OUT_C);
}
```

The number behind the **repeat** statement, between parentheses, indicates how often something must be repeated. The statements that must be repeated are put between brackets, just like the statements in a task. Note that, in the above program, we also indent the statements. This is not necessary, but it makes the program more readable.

As a final example, let us make the robot drive 10 times in a square. Here is the program:

```
#define MOVE_TIME    100
#define TURN_TIME     85

task main()
{
  repeat(10)
  {
    repeat(4)
    {
```

```
      OnFwd(OUT_A+OUT_C);
      Wait(MOVE_TIME);
      OnRev(OUT_C);
      Wait(TURN_TIME);
     }
   }
  Off(OUT_A+OUT_C);
 }
```

There is now one repeat statement inside the other. We call this a "nested" repeat statement. You can nest repeat statements as much as you like. Take a careful look at the brackets and the indentation used in the program. The task starts at the first bracket and ends at the last. The first repeat statement starts at the second bracket and ends at the fifth. The second, nested repeat statement starts at the third bracket and ends at the fourth. As you see the brackets always come in pairs, and the piece between the brackets we indent.

Adding Comment

To make your program even more readable, it is good to add some comment to it. Whenever you put // on a line, the rest of that line is ignored and can be used for comments. A long comment can be put between /* and */. Comments are syntax highlighted in the Bricx Command Center. The full program could look as follows:

```
/*  10 SQUARES

    by Mark Overmars

This program make the robot run 10 squares
*/

#define MOVE_TIME    100     // Time for a straight move
#define TURN_TIME     85     // Time for turning 90 degrees

task main()
{
  repeat(10)                 // Make 10 squares
  {
    repeat(4)
    {
      OnFwd(OUT_A+OUT_C);
      Wait(MOVE_TIME);
      OnRev(OUT_C);
      Wait(TURN_TIME);
    }
  }
  Off(OUT_A+OUT_C);          // Now turn the motors off
}
```

Using Variables

Variables form a very important aspect of every programming language. Variables are memory locations in which we can store a value. We can use that value at different places and we can change it. Let me describe the use of variables using an example.

Moving in a Spiral

Assume we want to adapt the above program in such a way that the robot drives in a spiral. This can be achieved by making the time we sleep larger for each next straight movement. That is, we want to increase the value of MOVE_TIME each time. But how can we do this? MOVE_TIME is a constant and constants cannot be changed. We need a variable instead. Variables can easily be defined in NQC. You can have 32 of these, and you can give each of them a separate name. Here is the spiral program.

```
#define TURN_TIME     85

int move_time;              // define a variable

task main()
{
  move_time = 20;           // set the initial value
  repeat(50)
  {
    OnFwd(OUT_A+OUT_C);
    Wait(move_time);        // use the variable for sleeping
    OnRev(OUT_C);
    Wait(TURN_TIME);
    move_time += 5;         // increase the variable
  }
  Off(OUT_A+OUT_C);
}
```

The interesting lines are indicated with the comments. First we define a variable by typing the keyword **int** followed by a name we choose. (Normally we use lower-case letters for variable names and uppercase letters for constants, but this is not necessary.) The name must start with a letter but can contain digits and the underscore sign. No other symbols are allowed. (The same applied to constants, task names, *etc.*) The strange word **int** stands for integer. Only integer numbers can be stored in it. In the second interesting line we assign the value 20 to the variable. From this moment on, whenever you use the variable, it stands for 20. Now follows the repeat loop in which we use the variable to indicate the time to sleep and, at the end of the loop we increase the value of the variable with 5. So the first time the robot sleeps 20 ticks, the second time 25, the third time 30, *etc.*

Besides adding values to a variable we can also multiply a variable with a number using *=, subtract using -= and divide using /=. You can also add one variable to the other, and write down more complicated expressions. Here are some examples:

```
int aaa;
int bbb, ccc;
task main()
{
  aaa = 10;
  bbb = 20 * 5;
  ccc = bbb;
  ccc /= aaa;
  ccc -= 5;
  aaa = 10 * (ccc + 3); // aaa is now equal to 80
}
```

Note on the first two lines that we can define multiple variables in one line. We could also have combined all three of them in one line.

Random numbers

In all the above programs we defined exactly what the robot was supposed to do. But things get a lot more interesting when the robot is going to do things that we don't know. We want some randomness in the motions. In NQC you can create random numbers. The following program uses this to let the robot drive around in a random way. It constantly drives forwards for a random amount of time and then makes a random turn.

```
int move_time, turn_time;

task main()
{
  while(true)
  {
    move_time = Random(60);
    turn_time = Random(40);
    OnFwd(OUT_A+OUT_C);
    Wait(move_time);
    OnRev(OUT_A);
    Wait(turn_time);
  }
}
```

The program defines two variables, and then assigns random numbers to them. `Random(60)` means a random number between 0 and 60 (it can also be 0 or 60). Each time the numbers will be different. (Note that we could avoid the use of the variables by writing *e.g.* `Wait(Random(60))`.)

You also see a new type of loop here. Rather that using the repeat statement we wrote **while**(**true**). The while statement repeats the statements below it as long as the condition between the parentheses is true. The special word **true** is always true, so the statements between the brackets are repeated forever, just as we want.

Control Structures

We saw the repeat and while statements. These statements control the way the other statements in the program are executed. They are called "control structures".

The if Statement

Sometimes you want that a particular part of your program is only executed in certain situations. In this case the if statement is used. Let me give an example. We will again change the program we have been working with so far, but with a new twist. We want the robot to drive along a straight line and then either make a left or a right turn. To do this we need random numbers again. We pick a random number between 0 and 1, that is, it is either 0 or 1. If the number is 0 we make a right turn; otherwise we make a left turn. Here is the program:

```
#define MOVE_TIME    100
#define TURN_TIME     85

task main()
{
  while(true)
  {
    OnFwd(OUT_A+OUT_C);
    Wait(MOVE_TIME);
    if (Random(1) == 0)
    {
      OnRev(OUT_C);
    }
    else
    {
      OnRev(OUT_A);
    }
    Wait(TURN_TIME);
  }
}
```

The if statement looks a bit like the while statement. If the condition between the parentheses is true the part between the brackets is executed. Otherwise, the part between the brackets after the word **else** is executed. Let us look a bit better at the condition we use. It reads `Random(1) == 0`. This means that `Random(1)` must be equal to 0 to make the condition true. You might wonder why we use == rather than =. The reason is to distinguish it from the statement that put a value in a variable. You can compare values in different ways. Here are the most important ones:

== equal to
< smaller than
<= smaller than or equal to
> larger than
>= larger than or equal to
!= not equal to

You can combine conditions use &&, which means "and", or | |, which means "or". Here are some examples of conditions:

true	always true
false	never true
`ttt != 3`	true when ttt is not equal to 3
`(ttt >= 5) && (ttt <= 10)`	true when ttt lies between 5 and 10
`(aaa == 10) || (bbb == 10)`	true if either aaa or bbb (or both) are equal to 10

Note that the if statement has two parts. The part immediately after the condition, which is executed when the condition is true, and the part after the else, which is executed when the condition is false. The keyword else and the part after it are optional. So you can leave them away if there is nothing to do when the condition is false.

The Do Statement

There is another control structure, the do statement. It has the following form:

```
do
{
   statements;
}
while (condition);
```

The statements between the brackets after the do part are executed as long as the condition is true. The condition has the same form as in the if statement described above. Here is an example of a program. The robot runs around randomly for 20 seconds and then stops.

```
int move_time, turn_time, total_time;

task main()
{
  total_time = 0;
  do
  {
    move_time = Random(100);
    turn_time = Random(100);
    OnFwd(OUT_A+OUT_C);
    Wait(move_time);
    OnRev(OUT_C);
    Wait(turn_time);
    total_time += move_time; total_time += turn_time;
  }
  while (total_time < 2000);
  Off(OUT_A+OUT_C);
}
```

Note in this example that we placed two statements on one line. This is allowed. You can place as many statements on a line as you like (as long as there are semicolons in between). But for readability of the program this is often not a good idea.

Note also that the do statement behaves almost the same as the while statement. But in the while statement the condition is tested before executing the statements, while in the do statement the condition is tested at the end. For the while statement, the statements might never be executed, but for the do statement they are executed at least once.

Sensors

One of the nice aspects of the Lego robots is that you can connect sensors to them and that you can make the robot react to the sensors. Before I can show how to do this we must change the robot a bit by adding a sensor. To this end, build the sensor construction on page 28 of the constructopedia. You might want to make it slightly wider, such that your robot looks as follows:

Fig. : Connect the sensor to input 1 on the RCX.

Waiting for a Sensor

Let us start with a very simple program in which the robot drives forwards until it hits something. Here it is:

```
task main()
{
    SetSensor(SENSOR_1,SENSOR_TOUCH);
    OnFwd(OUT_A+OUT_C);
    until (SENSOR_1 == 1);
    Off(OUT_A+OUT_C);
}
```

There are two important lines here. The first line of the program tells the robot what type of sensor we use. SENSOR_1 is the number of the input to which we connected the sensor. The other two sensor inputs are called SENSOR_2 and SENSOR_3. SENSOR_TOUCH indicates that this is a touch sensor. For the light sensor we would use SENSOR_LIGHT. After we specified the type of the sensor, the program switches on both motors and the robot starts moving forwards. The next statement is a very useful construction. It waits until the condition between the brackets is true. This condition says that the value of the sensor SENSOR_1 must be 1, which means that the sensor is pressed. As long as the sensor is not pressed, the value is 0. So this statement waits until the sensor is pressed. Then we switch off the motors and the task is finished.

Acting on a Touch Sensor

Let us now try to make the robot avoid obstacles. Whenever the robot hits an object, we let it move back a bit, make a turn, and then continue. Here is the program:

```
task main()
{
  SetSensor(SENSOR_1,SENSOR_TOUCH);
  OnFwd(OUT_A+OUT_C);
  while (true)
  {
    if (SENSOR_1 == 1)
    {
      OnRev(OUT_A+OUT_C); Wait(30);
      OnFwd(OUT_A); Wait(30);
      OnFwd(OUT_A+OUT_C);
    }
  }
}
```

As in the previous example, we first indicate the type of the sensor. Next the robot starts moving forwards. In the infinite while loop we constantly test whether the sensor is touched and, if so, move back for 1/3 of a second, turn right for 1/3 of a second, and then continue forwards again.

Light Sensors

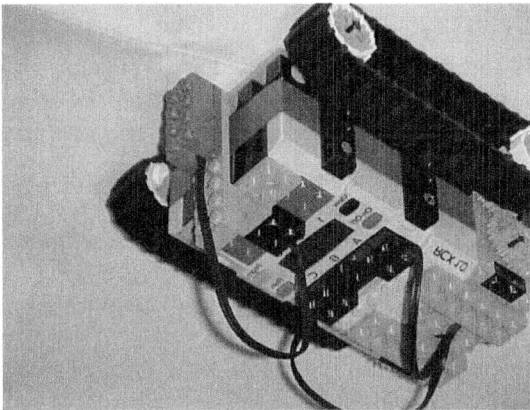

Besides touch sensors, you also get a light sensor with your MindStorms system. The light sensor measures the amount of light in a particular direction. The light sensor also emits light. In this way it is possible to point the light sensor in a particular direction and make a distinction between the intensity of the object in that direction. This is in particular useful when trying to make a robot follow a line on the floor. This is what we are going to do in the next example. We first need to attach the light sensor to the robot such that it is in the middle of the robot, at the front, and points downwards. Connect it to input 2. For example, make a construction as follows:

We also need the race track that comes with the RIS kit (This big piece of paper with the black track on it.) The idea now is that the robot makes sure that the light sensor stays above the track. Whenever the intensity of the light goes up, the light sensor is off the track and we need to adapt the direction. Here is a very simple program for this that only works if we travel around the track in clockwise direction.

```
#define THRESHOLD 40

task main()
{
  SetSensor(SENSOR_2, SENSOR_LIGHT);
  OnFwd(OUT_A+OUT_C);
  while (true)
  {
    if (SENSOR_2 > THRESHOLD)
    {
      OnRev(OUT_C);
      until (SENSOR_2 <= THRESHOLD);
      OnFwd(OUT_A+OUT_C);
    }
  }
}
```

The program first indicates that sensor 2 is a light sensor. Next it sets the robot to move forwards and goes into an infinite loop. Whenever the light value is bigger than 40 (we use a constant here such that this can be adapted easily, because it depends a lot on the surrounding light) we reverse one motor and wait till we are on the track again.

As you will see when you execute the program, the motion is not very smooth. Try adding a `Wait(10)` command before the **until** command to make the robot move better. Note that the program does not work for moving counter-clockwise. To enable motion along arbitrary path a much more complicated program is required.

Tasks and Subroutines

Up to now all our programs consisted of just one task. But NQC programs can have multiple tasks. It is also possible to put pieces of code in so-called subroutines that you can use at different places in your program. Using tasks and subroutines makes your programs easier to understand and more compact.

Tasks

An NQC program consists of at most 10 tasks. Each task has a name. One task must have the name `main`, and this task will be executed. The other tasks will only be executed when a running tasks tells them to be executed using a start command. From this moment on both tasks are running simultaneously (so the first task continues running). A running task can also stop another running task by using the stop command. Later this task can be restarted again, but it will start from the beginning; not from the place where it was stopped.

Let me demonstrate the use of tasks. Put your touch sensor again on your robot. We want to make a program in which the robot drives around in squares, like before. But when it hits an obstacle it should react to it. It is difficult to do this in one task, because the robot must do two things at the same moment: drive around (that is, switching on and off motors at the right moments) and watch for sensors. So it is better to use two tasks for this, one task that drives the squares; the other that reacts to the sensors. Here is the program.

```
task main()
{
  SetSensor(SENSOR_1,SENSOR_TOUCH);
  start check_sensors;
  start move_square;
}

task move_square()
{
  while (true)
  {
    OnFwd(OUT_A+OUT_C); Wait(100);
    OnRev(OUT_C); Wait(85);
  }
}

task check_sensors()
{
  while (true)
  {
    if (SENSOR_1 == 1)
    {
      stop move_square;
      OnRev(OUT_A+OUT_C); Wait(50);
      OnFwd(OUT_A); Wait(85);
      start move_square;
    }
  }
}
```

The main task just sets the sensor type and then starts both other tasks. After this, task main is finished. Task move_square moves the robot forever in squares. Task check_sensors checks whether the touch sensor is pushed. If so it takes the following actions: First of all it stops task move_square. This is very important. check_sensors now takes control over the motions of the robot. Next it moves the robot back a bit and makes it turn. Then it can start move_square again to let the robot again drive in squares.

It is very important to remember that tasks that you start are running at the same moment. This can lead to unexpected results.

Subroutines

Sometimes you need the same piece of code at multiple places in your program. In this case you can put the piece of code in a subroutine and give it a name. Now you can execute this piece of code by simply calling its name from within a task. NQC (or actually the RCX) allows for at most 8 subroutines. Let us look at an example.

```
sub turn_around()
{
  OnRev(OUT_C); Wait(340);
  OnFwd(OUT_A+OUT_C);
}

task main()
{
  OnFwd(OUT_A+OUT_C);
  Wait(100);
  turn_around();
  Wait(200);
  turn_around();
  Wait(100);
  turn_around();
  Off(OUT_A+OUT_C);
}
```

In this program we have defined a subroutine that makes the robot rotate around its center. The main task calls the subroutine three times. Note that we call the subroutine by writing down its name with parentheses behind it. So it looks the same as many of the commands we have seen. Only there are no parameters, so there is nothing between the parentheses.

Some warnings are in place here. Subroutines are a bit weird. For example, subroutines cannot be called from other subroutines. Subroutines can be called from different tasks but this is not encouraged. It very easily leads to problems because the same subroutine might actually be run twice at the same moment by different tasks. This tends to give unwanted effects. Also, when calling a subroutine from different tasks, due to a limitation in the RCX firmware, you cannot use complicated expressions anymore. So, unless you know precisely what you are doing, *don't call a subroutine from different tasks!*

Inline Functions

As indicated above, subroutines cause certain problems. The nice part is that they are stored only once in the RCX. This saves memory and, because the RCX does not have so much free memory, this is useful. But when subroutines are short, better use inline functions instead. These are not stored separately but copied at each place they are used. This costs more memory but problems like the ones with using complicated expressions, are no longer present. Also there is no limit on the number of inline functions.

Defining and calling inline functions goes exactly the same way as with subroutines. Only use the keyword **void** rather than **sub**. So the above example, using inline functions, looks as follows:

```
void turn_around()
{
  OnRev(OUT_C); Wait(340);
  OnFwd(OUT_A+OUT_C);
}

task main()
{
  OnFwd(OUT_A+OUT_C);
  Wait(100);
  turn_around();
  Wait(200);
  turn_around();
  Wait(100);
  turn_around();
  Off(OUT_A+OUT_C);
}
```

Inline functions have another advantage over subroutines. They can have arguments. Arguments can be used to pass a value for certain variables to an inline function. For example, assume, in the above example, we can make the time to turn an argument of the function, as in the following examples:

```
void turn_around(int turntime)
{
  OnRev(OUT_C); Wait(turntime);
  OnFwd(OUT_A+OUT_C);
}
task main()
{
  OnFwd(OUT_A+OUT_C);
  Wait(100);
  turn_around(200);
  Wait(200);
  turn_around(50);
  Wait(100);
  turn_around(300);
  Off(OUT_A+OUT_C);
}
```

Note that in the parenthesis behind the name of the inline function we specify the argument(s) of the function. In this case we indicate that the argument is an integer (there are some other choices) and that its name is turntime. When there are more arguments, you must separate them with commas.

Defining Macros

There is yet another way to give small pieces of code a name. You can define macros in NQC (not to be confused with the macros in Bricx Command Center). We have seen before that we can define constants, using #define, by giving them a name. But actually we can define any piece of code. Here is the same program again but now using a macro for turning around.

```
#define turn_around OnRev(OUT_C);Wait(340);OnFwd(OUT_A+OUT_C);
task main()
{
  OnFwd(OUT_A+OUT_C);
  Wait(100);
  turn_around;
  Wait(200);
  turn_around;
  Wait(100);
  turn_around;
  Off(OUT_A+OUT_C);
}
```

After the #define statement the word turn_around stands for the text behind it. Now wherever you type turn_around, this is replaced by this text. Note that the text should be on one line. (Actually there are ways of putting a #define statement on multiple lines, but this is not recommended.)

Define statements are actually a lot more powerful. They can also have arguments. For example, we can put the time to turn as an argument in the statement. Here is an example in which we define four macro's; one to move forwards, one to move backwards, one to turn left and one to turn right. Each has two arguments: the speed and the time.

```
    #define turn_right(s,t)   SetPower(OUT_A+OUT_C,s);OnF
wd(OUT_A);OnRev(OUT_C);Wait(t);
    #define turn_left(s,t)    SetPower(OUT_A+OUT_C,s);OnR
ev(OUT_A);OnFwd(OUT_C);Wait(t);
    #define forwards(s,t)   SetPower(OUT_A+OUT_C,s);OnFwd(OUT_
A+OUT_C);Wait(t);
    #define  backwards(s,t)        SetPower(OUT_
A+OUT_C,s);OnRev(OUT_A+OUT_C);Wait(t);

  task main()
  {
    forwards(3,200);
    turn_left(7,85);
    forwards(7,100);
    backwards(7,200);
    forwards(7,100);
    turn_right(7,85);
    forwards(3,200);
    Off(OUT_A+OUT_C);
  }
```

It is very useful to define such macros. It makes your code more compact and readable. Also, you can more easily change your code when you *e.g.* change the connections to the motors.

Making Music

The RCX has a built-in speaker that can make sounds and even play simple pieces of music. This is in particular useful when you want to make the RCX tell you that something is happening. But it can also be funny to have the robot make music while it runs around.

Built-in Sounds

There are six built-in sounds in the RCX, numbered from 0 to 5. They sound as follows:

0   Key click
1   Beep beep
2   Decreasing frequency sweep
3   Increasing frequency sweep
4   'Buhhh' Error sound
5   Fast increasing sweep

You can play them using the commands `PlaySound()`. Here is a small program that plays all of them.

```
task main()
{
  PlaySound(0); Wait(100);
  PlaySound(1); Wait(100);
  PlaySound(2); Wait(100);
  PlaySound(3); Wait(100);
  PlaySound(4); Wait(100);
  PlaySound(5); Wait(100);
}
```

You might wonder why there are these wait commands. The reason is that the command that plays the sound does not wait for it to finish. It immediately executes the next command. The RCX has a little buffer in which it can store some sounds but after a while this buffer get full and sounds get lost.

Playing Music

For more interesting music, NQC has the command `PlayTone()`. It has two arguments. The first is the frequency, and the second the duration (in ticks of 1/100h of a second, like in the wait command). Here is a table of useful frequencies:

Sound	1	2	3	4	5	6	7	8
G#	52	104	208	415	831	1661	3322	
G	49	98	196	392	784	1568	3136	
F#	46	92	185	370	740	1480	2960	
F	44	87	175	349	698	1397	2794	
E	41	82	165	330	659	1319	2637	
D#	39	78	156	311	622	1245	2489	
D	37	73	147	294	587	1175	2349	
C#	35	69	139	277	554	1109	2217	
C	33	65	131	262	523	1047	2093	4186
B	31	62	123	247	494	988	1976	3951
A#	29	58	117	233	466	932	1865	3729
A	28	55	110	220	440	880	1760	3520

As we noted above for sounds, also here the RCX does not wait for the note to finish. So if you use a lot in a row better add (slightly longer) wait commands in between. Here is an example:

```
task main()
{
  PlayTone(262,40);  Wait(50);
  PlayTone(294,40);  Wait(50);
  PlayTone(330,40);  Wait(50);
  PlayTone(294,40);  Wait(50);
  PlayTone(262,160); Wait(200);
}
```

You can create pieces of music very easily using the Brick Piano that is part of the Bricx Command Center.

If you want to have the RCX play music while driving around, better use a separate task for it. Here you have an example of a rather stupid program where the RCX drives back and forth, constantly making music.

```
task music()
{
  while (true)
  {
    PlayTone(262,40);  Wait(50);
    PlayTone(294,40);  Wait(50);
    PlayTone(330,40);  Wait(50);
    PlayTone(294,40);  Wait(50);
  }
}

task main()
{
  start music;
  while(true)
  {
    OnFwd(OUT_A+OUT_C); Wait(300);
    OnRev(OUT_A+OUT_C); Wait(300);
  }
}
```

More about Motors

There are a number of additional motor commands that you can use to control the motors more precisely.

Stopping Gently

When you use the `Off()` command, the motor stops immediately, using the brake. In NQC it is also possible to stop the motors in a more gentle way, not using the brake. For this you use the `Float()` command. Sometimes this is better for your robot task. Here is an example. First the robot stops using the brakes; next without using the brakes. Note the difference. (Actually the difference is very small for this particular robot. But it makes a big difference for some other robots.)

```
task main()
{
    OnFwd(OUT_A+OUT_C);
    Wait(200);
    Off(OUT_A+OUT_C);
    Wait(100);
    OnFwd(OUT_A+OUT_C);
    Wait(200);
    Float(OUT_A+OUT_C);
}
```

Advanced Commands

The command `OnFwd()` actually does two things: it switches the motor on and it sets the direction to forwards. The command `OnRev()` also does two things: it switches the motor on and sets the direction to reverse. NQC also has commands to do these two things separately. If you only want to change one of the two things, it is more efficient to use these separate commands; it uses less memory in the RCX, it is faster, and it can result in smoother motions. The two separate commands are `SetDirection()` that sets the direction (`OUT_FWD`, `OUT_REV` or `OUT_TOGGLE` which flips the current direction) and `SetOutput()` that sets the mode (`OUT_ON`, `OUT_OFF` or `OUT_FLOAT`). Here is a simple program that makes the robot drive forwards, backwards and forwards again.

```
task main()
{
    SetPower(OUT_A+OUT_C,7);
    SetDirection(OUT_A+OUT_C,OUT_FWD);
    SetOutput(OUT_A+OUT_C,OUT_ON);
    Wait(200);
    SetDirection(OUT_A+OUT_C,OUT_REV);
    Wait(200);
    SetDirection(OUT_A+OUT_C,OUT_TOGGLE);
    Wait(200);
    SetOutput(OUT_A+OUT_C,OUT_FLOAT);
}
```

Note that, at the start of every program, all motors are set in forward direction and the speed is set to 7. So in the above example, the first two commands are not necessary.

There are a number of other motor commands, which are shortcuts for combinations of the commands above. Here is a complete list:

`On('motors')`	Switches the motors on
`Off('motors')`	Switches the motors off
`Float('motors')`	Switches the motors of smoothly
`Fwd('motors')`	Switches the motors forward (but does not make them drive)
`Rev('motors')`	Switches the motors backwards (but does not make them drive)
`Toggle('motors')`	Toggles the direction of the motors (forward to backwards and back)
`OnFwd('motors')`	Switches the motors forward and turns them on
`OnRev('motors')`	Switches the motors backwards and turns them on
`OnFor('motors','ticks')`	Switches the motors on for ticks time
`SetOutput('motors','mode')`	Sets the output mode (OUT_ON, OUT_OFF or OUT_FLOAT)
`SetDirection('motors','dir')`	Sets the output direction (OUT_FWD, OUT_REV or OUT_TOGGLE)
`SetPower('motors','power')`	Sets the output power (0-9)

Varying Motor Speed

As you probably noticed, changing the speed of the motors does not have much effect. The reason is that you are mainly changing the torque, not the speed. You will only see an effect when the motor has a heavy load. And even then, the difference between 2 and 7 is very small. If you want to have better effects the trick is to turn the motors on and off in rapid succession. Here is a simple program that does this. It has one task, called run_motor that drives the motors. It constantly checks the variable speed to see what the current speed is. Positive is forwards, negative backwards. It sets the motors in the right direction and then waits for some time, depending on speed, before switching the motors off again. The main task simply sets speeds and waits.

```
int speed, __speed;

task run_motor()
{
  while (true)
  {
    __speed = speed;
    if (__speed > 0) {OnFwd(OUT_A+OUT_C);}
    if (__speed < 0) {OnRev(OUT_A+OUT_C); __speed = -__speed;}
    Wait(__speed);
    Off(OUT_A+OUT_C);
  }
}

task main()
{
  speed = 0;
  start run_motor;
  speed = 1;    Wait(200);
  speed = -10;  Wait(200);
  speed = 5;    Wait(200);
  speed = -2;   Wait(200);
  stop run_motor;
  Off(OUT_A+OUT_C);
}
```

This program can be made much more powerful, allowing for rotations, and also possibly incorporating a waiting time after the Off() command. Experiment yourself.

More about Sensors

The basic aspects of using sensors. But there is a lot more you can do with sensors. The difference between sensor mode and sensor type, we will see how to use the rotation sensor (a type of sensor that is not provided with the RIS but can be bought separately and is very useful), and we will see some tricks to use more than three sensors and to make a proximity sensor.

Sensor Mode and Type

The SetSensor() command that we saw before does actually two things: it sets the type of the sensor, and it sets the mode in which the sensor operates. By setting the mode and type of the a sensor separately, you can control the behavior of the sensor more precisely, which is useful for particular applications.

The type of the sensor is set with the command SetSensorType(). There are four different types: SENSOR_TYPE_TOUCH, which is the touch sensor, SENSOR_TYPE_LIGHT, which is the light sensor, SENSOR_TYPE_TEMPERATURE, which is the temperature sensor (this type of sensor is not part of the RIS but can be bought separately), and SENSOR_TYPE_ROTATION, which is the rotation sensor (also not part of the RIS but available separately). Setting the type sensor is in particular important to indicate whether the sensor needs power (like *e.g.* for the light of the light sensor). I know of no uses for setting a sensor to a different type than it actually is.

The mode of the sensor is set with the command SetSensorMode(). There are eight different modes. The most important one is SENSOR_MODE_RAW. In this mode, the value you get when checking the sensor is a number between 0 and 1023. It is the raw value produced by the sensor. What it means depends on the actual sensor. For example, for a touch sensor, when the sensor is not pushed the value is close to 1023. When it is fully pushed, it is close to 50. When it is pushed partially the value ranges between 50 and 1000. So if you set a touch sensor to raw mode you can actually find out whether it is touched partially. When the sensor is a light sensor, the value ranges from about 300 (very light) to 800 (very dark). This gives a much more precise value than using the SetSensor() command.

The second sensor mode is SENSOR_MODE_BOOL. In this mode the value is 0 or 1. When the raw value is above about 550 the value is 0, otherwise it is 1. SENSOR_MODE_BOOL is the default mode for a touch sensor. The modes SENSOR_MODE_CELSIUS and SENSOR_MODE_FAHRENHEIT are useful with temperature sensors only and give the temperature in the indicated way. SENSOR_MODE_PERCENT turns the raw value into a value between 0 and 100. Every raw value of 400 or lower is mapped to 100 percent. If the raw value gets higher, the percentage slowly goes down to 0. SENSOR_MODE_PERCENT is the default mode for a light sensor. SENSOR_MODE_ROTATION seems to be useful only for the rotation sensor.

There are two other interesting modes: SENSOR_MODE_EDGE and SENSOR_MODE_PULSE. They count transitions, that is changes from a low to a high raw value or opposite. For example, when you touch a touch sensor this causes a transition from high to low raw value. When you release it you get a transition the other direction. When you set the sensor mode to SENSOR_MODE_PULSE, only transitions from low to high are counted. So each touch and release of the touch sensor counts for one. When you set the sensor mode to SENSOR_MODE_EDGE, both transitions are counted. So each touch and release of the touch sensor counts for two. So you can use this to count how often a touch sensor is pushed. Or you can use it in combination with a light sensor to count how often a (strong) lamp is switched on and off. Of course, when you are counting things, you should be able to set the counter back to 0. For this you use the command ClearSensor(). It clears the counter for the indicated sensor(s).

Let us look at an example. The following program uses a touch sensor to steer the robot. Connect the touch sensor with a long wire to input one. If touch the sensor quickly twice the robot moves forwards. It you touch it once it stops moving.

```
task main()
{
  SetSensorType(SENSOR_1,SENSOR_TYPE_TOUCH);
  SetSensorMode(SENSOR_1,SENSOR_MODE_PULSE);
  while(true)
  {
    ClearSensor(SENSOR_1);
    until (SENSOR_1 >0);
    Wait(100);
    if (SENSOR_1 == 1) {Off(OUT_A+OUT_C);}
    if (SENSOR_1 == 2) {OnFwd(OUT_A+OUT_C);}
  }
}
```

Note that we first set the type of the sensor and then the mode. It seems that this is essential because changing the type also effects the mode.

The Rotation Sensor

The rotation sensor is a very useful type of sensor that is unfortunately not part of the standard RIS. It can though be bought separately from Lego. The rotation sensor contains a hole through which you can put an axle. The rotation sensor measures the amount the axle is rotated. One full rotation of the axle is 16 steps (or –16 if you rotate it the other way). Rotation sensors are very useful to make the robot make precisely controlled movements. You can make an axle move the exact amount you want. If you need finer control than 16 step, you can always use gears to connect it to an axle that moves faster, and use that one for counting steps.

One standard application is to have two rotation sensors connected to the two wheels of the robot that you control with the two motors. For a straight movement you want both wheels to turn equally fast. Unfortunately, the motors normally don't run at exactly the same speed. Using the rotation sensors you can see that one wheel turns faster. You can then temporarily stop that motor (best using Float()) until both sensors give the same value again. The following program does this. It simply lets the robot drive in a straight line. To use it, change your robot by connecting the two rotation sensors to the two wheels. Connect the sensors to input 1 and 3.

```
task main()
{
  SetSensor(SENSOR_1,SENSOR_ROTATION); ClearSensor(SENSOR_1);
  SetSensor(SENSOR_3,SENSOR_ROTATION); ClearSensor(SENSOR_3);
  while (true)
  {
    if (SENSOR_1 < SENSOR_3)
      {OnFwd(OUT_A); Float(OUT_C);}
    else if (SENSOR_1 > SENSOR_3)
      {OnFwd(OUT_C); Float(OUT_A);}
    else
      {OnFwd(OUT_A+OUT_C);}
  }
}
```

The program first indicates that both sensors are rotation sensors, and resets the values to zero. Next it start an infinite loop. In the loop we check whether the two sensor readings are equal. If they are the robot simply moves forwards. If one is larger, the correct motor is stopped until both readings are again equal.

Clearly this is only a very simple program. You can extend this to make the robot drive exact distances, or to let it make very precise turns.

Putting Multiple Sensors on One Input

The RCX has only three inputs so you can connect only three sensors to it. When you want to make more complicated robots (and you bought some extra sensors) this might not be enough for you. Fortunately, with some tricks, you can connect two (or even more) sensors to one input.

The easiest is to connect two touch sensors to one input. If one of them (or both) is touched, the value is 1, otherwise it is 0. You cannot distinguish the two but sometimes this is not necessary. For example, when you put one touch sensor at the front and one at the back of the robot, you know which one is touched based on the direction the robot is driving in. But you can also set the mode of the input to raw. Now you can get a lot more information. If you are lucky, the value when the sensor is pressed is not the same for both sensors. If this is the case you can actually distinguish between the two sensors. And when both are pressed you get a much lower value (around 30) so you can also detect this.

You can also connect a touch sensor and a light sensor to one input. Set the type to light (otherwise the light sensor won't work). Set the mode to raw. In this case, when the touch sensor is pushed you get a raw value below 100. If it is not pushed you get the value of the light sensor which is never below 100. The following program uses this idea. The robot must be equipped with a light sensor pointing down, and a bumper at the front connected to a touch sensor. Connect both of them to input 1. The robot will drive around randomly within a light area.

When the light sensor sees a dark line (raw value > 750) it goes back a bit. When the touch sensor touches something (raw value below 100) it does the same. Here is the program:

```
int ttt,tt2;
task moverandom()
{
   while (true)
   {
      ttt = Random(50) + 40;
      tt2 = Random(1);
      if (tt2 > 0)
         { OnRev(OUT_A); OnFwd(OUT_C); Wait(ttt); }
      else
         { OnRev(OUT_C); OnFwd(OUT_A);Wait(ttt); }
      ttt = Random(150) + 50;
      OnFwd(OUT_A+OUT_C);Wait(ttt);
   }
}

task main()
{
   start moverandom;
   SetSensorType(SENSOR_1,SENSOR_TYPE_LIGHT);
   SetSensorMode(SENSOR_1,SENSOR_MODE_RAW);
   while (true)
   {
      if ((SENSOR_1 < 100) || (SENSOR_1 > 750))
      {
         stop moverandom;
         OnRev(OUT_A+OUT_C);Wait(30);
         start moverandom;
      }
   }
}
```

I hope the program is clear. There are two tasks. Task moverandom makes the robot move around in a random way. The main task first starts moverandom, sets the sensor and then waits for something to happen. If the sensor reading gets too low (touching) or too high (out of the white area) it stops the random moves, backs up a little, and start the random moves again.

It is also possible to connect two light sensors to the same input. The raw value is in some way related to the combined amount of light received by the two sensors. But this is rather unclear and seems hard to use. Connecting other sensors with rotation or temperature sensors seems not to be useful.

Making a Proximity Sensor

Using touch sensors, your robot can react when it hits something. But it would be a lot nicer when the robot could react just before it hits something. It should know that it is near to some obstacle. Unfortunately there are no sensors for this available. There is though a trick we can use for this. The robot has an infra-red port with which it can communicate with the computer, or with other robots. It turns out that the light sensor that comes with the robot is very sensitive to infra-red light. We can build a proximity sensor based on this. The idea is that one tasks sends out infra-red messages. Another task measures fluctuations in the light intensity that is reflected from objects. The higher the fluctuation, the closer we are to an object.

To use this idea, place the light sensor above the infra-red port on the robot, pointing forwards. In this way it only measures reflected infra-red light. Connect it to input 2. We use raw mode for the light sensor to see the fluctuations as good as possible. Here is a simple program that lets the robot run forwards until it gets near to an object and then makes a 90 degree turn to the right.

```
int lastlevel;              // To store the previous level

task send_signal()
{
  while(true)
    {SendMessage(0); Wait(10);}
}
task check_signal()
{
  while(true)
  {
    lastlevel = SENSOR_2;
    if(SENSOR_2 > lastlevel + 200)
      {OnRev(OUT_C); Wait(85); OnFwd(OUT_A+OUT_C);}
  }
}
task main()
{
  SetSensorType(SENSOR_2, SENSOR_TYPE_LIGHT);
  SetSensorMode(SENSOR_2, SENSOR_MODE_RAW);
  OnFwd(OUT_A+OUT_C);
  start send_signal;
  start check_signal;
}
```

The task `send_signal` send out 10 IR signals every seconds, using the command `SendMessage(0)`. The task `check_signal` repeatedly saves the value of the light sensor. Then it checks whether it (slightly later) has become at least 200 higher, indicating a large fluctuation. If so, it lets the robot make a 90-degree turn to the right. The value of 200 is rather arbitrary. If you make it smaller, the robot turns further away from obstacles. If you make it larger, it gets closer to them. But this also depends on the type of material and the amount of light available in the room. You should experiment or use some more clever mechanism for learning the correct value.

A disadvantage of the technique is that it only works in one direction. You probably still need touch sensors at the sides to avoid collisions there. But the technique is very useful for robots that must drive around in mazes. Another disadvantage is that you cannot communicate from the computer to the robot because it will interfere with the infra-red commands send out by the robot. (Also the remote control on your television might not work.)

Parallel Tasks

As has been indicated before, tasks in NQC are executed simultaneously, or in parallel as people usually say. This is extremely useful. In enables you to watch sensors in one task while another task moves the robot around, and yet another task plays some music. But parallel tasks can also cause problems. One task can interfere with another.

A Wrong Program

Consider the following program. Here one task drives the robot around in squares (like we did so often before) and the second task checks for the touch sensor. When the sensor is touched, it moves a bit backwards, and makes a 90-degree turn.

This probably looks like a perfectly valid program. But if you execute it you will most likely find some unexpected behavior. Try the following: Make the robot touch something while it is turning. It will start going back, but immediately moves forwards again, hitting the obstacle. The reason for this is that the tasks may interfere. The following is happening. The robot is turning right, that is, the first task is in its second sleep statement. Now the robot hits the sensor. It start going backwards, but at that very moment, the main task is ready with sleeping and moves the robot forwards again; into the obstacle. The second task is sleeping at this moment so it won't notice the collision. This is clearly not the behavior we would like to see. The problem is that, while the second task is sleeping we did not realize that the first task was still running, and that its actions interfere with the actions of the second task.

```
task main()
{
  SetSensor(SENSOR_1,SENSOR_TOUCH);
  start check_sensors;
  while (true)
  {
    OnFwd(OUT_A+OUT_C); Wait(100);
    OnRev(OUT_C); Wait(85);
  }
}

task check_sensors()
{
  while (true)
  {
    if (SENSOR_1 == 1)
    {
      OnRev(OUT_A+OUT_C);
      Wait(50);
      OnFwd(OUT_A);
      Wait(85);
      OnFwd(OUT_C);
    }
  }
}
```

Stopping and Restarting Tasks

One way of solving this problem is to make sure that at any moment only one task is driving the robot.

The crux is that the check_sensors task only moves the robot after stopping the move_square task. So this task cannot interfere with the moving away from the obstacle. Once the backup procedure is finished, it starts move_square again.

Even though this is a good solution for the above problem, there is a problem. When we restart move_square, it starts again at the beginning. This is fine for our small task, but often this is not the required behavior. We would prefer to stop the task where it is and continue it later from that point. Unfortunately this cannot be done easily.

```
task main()
{
  SetSensor(SENSOR_1,SENSOR_TOUCH);
  start check_sensors;
  start move_square;
}

task move_square()
{
  while (true)
  {
    OnFwd(OUT_A+OUT_C); Wait(100);
    OnRev(OUT_C); Wait(85);
  }
}

task check_sensors()
{
  while (true)
  {
    if (SENSOR_1 == 1)
    {
      stop move_square;
      OnRev(OUT_A+OUT_C); Wait(50);
      OnFwd(OUT_A); Wait(85);
      start move_square;
    }
  }
}
```

Using Semaphores

A standard technique to solve this problem is to use a variable to indicate which task is in control of the motors. The other tasks are not allowed to drive the motors until the first task indicates, using the variable, that it is ready. Such a variable is often called a semaphore. Let sem be such a semaphore. We assume that a value of 0 indicates that no task is steering the motors. Now, whenever a task wants to do something with the motors it executes the following commands:

```
until (sem == 0);
sem = 1;
// Do something with the motors
sem = 0;
```

So we first wait till nobody needs the motors. Then we claim the control by setting sem to 1. Now we can control the motors. When we are done we set sem back to 0. Here you find the program above, implemented using a semaphore. When the touch sensor touches something, the semaphore is set and the backup procedure is performed. During this procedure the task move_square must wait. At the moment the back-up is ready, the semaphore is set to 0 and move_square can continue.

```
int sem;

task main()
{
   sem = 0;
   start move_square;
   SetSensor(SENSOR_1,SENSOR_TOUCH);
   while (true)
   {
     if (SENSOR_1 == 1)
     {
       until (sem == 0); sem = 1;
       OnRev(OUT_A+OUT_C); Wait(50);
       OnFwd(OUT_A); Wait(85);
       sem = 0;
     }
   }
}


task move_square()
{
   while (true)
   {
     until (sem == 0); sem = 1;
     OnFwd(OUT_A+OUT_C);
     sem = 0;
     Wait(100);
     until (sem == 0); sem = 1;
     OnRev(OUT_C);
     sem = 0;
     Wait(85);
   }
}
```

You could argue that it is not necessary in `move_square` to set the semaphore to 1 and back to 0. Still this is useful. The reason is that the `OnFwd()` command is in fact two commands. You don't want this command sequence to be interrupted by the other task.

Semaphores are very useful and, when you are writing complicated programs with parallel tasks, they are almost always required.

Communication between Robots

If you own more than one RCX. The robots can communicate with each other through the infra-red port. Using this you can have multiple robots collaborate (or fight with each other). Also you can build one big robot using two RCXs, such that you can have six motors and six sensors.

Communication between robots works, globally speaking, as follows. A robot can use the command `SendMessage()` to send a value (0-255) over the infra-red port. All other robots receive this message and store it. The program in a robot can ask for the value of the last message received using `Message()`. Based on this value the program can make the robot perform certain actions.

Giving Orders

Often, when you have two or more robots, one is the leader. We call him the *master*. The other robots are *slaves*. The master robot sends orders to the slaves and the slaves execute these. Sometimes the slaves might send information back to the master, for example the value of a sensor. So you need to write two programs, one for the master and one for the slave(s). From now on we assume that we have just one slave. Let us start with a very simple example. Here the slave can perform three different orders: move forwards, move backwards, and stop. Its program consists of a simple loop. In this loop it sets the value of the current message to 0 using the `ClearMessage()` command. Next it waits until the message becomes unequal to 0. Based on the value of the message it executes one of the three orders. Here is the program.

```
task main()            // SLAVE
{
  while (true)
  {
    ClearMessage();
    until (Message() != 0);
    if (Message() == 1) {OnFwd(OUT_A+OUT_C);}
    if (Message() == 2) {OnRev(OUT_A+OUT_C);}
    if (Message() == 3) {Off(OUT_A+OUT_C);}
  }
}
```

The master has an even simpler program. It simply send the messages corresponding to orders and then waits a bit. In the program below it orders the slave to move forwards, then, after two seconds, backwards, and then, again after two seconds, to stop.

```
task main()              // MASTER
{
   SendMessage(1); Wait(200);
   SendMessage(2); Wait(200);
   SendMessage(3);
}
```

After you have written these two program, you need to download them to the robots. Each program must go to one of the robots. Make sure you switch the other one off in the meantime. Now switch on both robots and start the programs: first the one in the slave and then the one in the master.

If you have multiple slaves, you have to download the slave program to each of them in turn. Now all slaves will perform exactly the same actions.

To let the robots communicate with each other we defined, what is called, a protocol: We decided that a 1 means to move forwards, a 2 to move backwards, and a 3 to stop. It is very important to carefully define such protocols, in particular when you are dealing with lots of communications. For example, when there are more slaves, you could define a protocol in which two numbers are sent (with a small sleep in between): the first number is the number of the slave, and the second is the actual order. The slave than first check the number and only perform the action if it is his number.

Electing a Leader

When dealing with multiple robots, each robot must have its own program. It would be much easier if we could download just one program to all robots. But then the question is: who is the master? The answer is easy: let the robots decide themselves. Let them elect a leader which the others will follow. But how do we do this? The idea is rather simple. We let each robot wait a random amount of time and then send a message. The one that sends a message first is the leader. This scheme might fail if two robots wait exactly the same amount of time but this is rather unlikely. Here is the program that does it:

```
task main()
{
  ClearMessage();
  Wait(200);                  // make sure all robots are on
  Wait(Random(400));          // wait between 0 and 4 seconds
  if (Message() > 0)          // somebody else was first
  {
    start slave;
  }
  else
  {
    SendMessage(1);           // I am the master now
    Wait(400);                // make sure everybody else knows
    start master;
  }
}

task master()
{
  SendMessage(1); Wait(200);
  SendMessage(2); Wait(200);
  SendMessage(3);
}

task slave()
{
  while (true)
  {
    ClearMessage();
    until (Message() != 0);
    if (Message() == 1) {OnFwd(OUT_A+OUT_C);}
    if (Message() == 2) {OnRev(OUT_A+OUT_C);}
    if (Message() == 3) {Off(OUT_A+OUT_C);}
  }
}
```

Download this program to all robots (one by one, not at the same moment). Start the robots at about the same moment and see what happens. One of them should take command and the other(s) should follow the orders. In rare occasions, none of them becomes the leader. As indicated above, this requires more careful protocols to solve.

Cautions

You have to be a bit careful when dealing with multiple robots. There are two problems: If two robots (or a robot and the computer) send information at the same time this might be lost. The second problem is that, when the computer sends a program to multiple robots at the same time, this causes problems.

Let us start with the second problem. When you download a program to the robot, the robot tells the computer whether it correctly receives (parts of) the program. The computer reacts on that by sending new pieces or by resending parts. When two robots are on, both will start telling the computer whether they correctly receive the program. The computer does not understand this (it does not know that there are two robots!). As a result, things go wrong and the program gets corrupted. The robots won't do the right things. *Always make sure that, while you are downloading programs, only one robot is on!*

The other problem is that only one robot can send a message at any moment. If two messages are being send at roughly the same moment, they might get lost. Also, a robot cannot send and receive messages at the same moment. This is no problem when only one robot sends messages (there is only one master) but otherwise it might be a serious problem. For example, you can imagine writing a program in which a slave sends a message when it bumps into something, such that the master can take action. But if the master sends an order at the same moment, the message will get lost. To solve this, it is important to define your communication protocol such that, in case a communication fails, this is corrected. For example, when the master sends a command, it should get an answer from the slave. If is does not get an answer soon enough, it resends the command. This would result in a piece of code that looks like this:

```
do
{
   SendMessage(1);
   ClearMessage();
   Wait(10);
}
while (Message() != 255);
```

Here 255 is used for the acknowledgement.

Sometimes, when you are dealing with multiple robots, you might want that only a robot that is very close by receives the signal. This can be achieved by adding the command SetTxPower(TX_POWER_LO) to the program of the master. In this case the IR signal send is very low and only a robot close by and facing the master will "hear" it. This is in particular useful when building one bigger robot out of two RCXs. Use `SetTxPower(TX_POWER_HI)` to set the robot again in long range transmission mode.

More Commands

NQC has a number of additional commands.

Timers

The RCX has four built-in timers. These timers tick in increments of 1/10 of a second. The timers are numbered from 0 to 3. You can reset the value of a timer with the command `ClearTimer()` and get the current value of the timer with `Timer()`. Here is an example of the use of a timer. The following program lets the robot drive sort of random for 20 seconds.

```
task main()
{
   ClearTimer(0);
   do
   {
     OnFwd(OUT_A+OUT_C);
     Wait(Random(100));
     OnRev(OUT_C);
     Wait(Random(100));
   }
   while (Timer(0)<200);
   Off(OUT_A+OUT_C);
}
```

Timers are very useful as a replacement for a `Wait()` command. You can sleep for a particular amount of time by resetting a timer and then waiting till it reaches a particular value. But you can also react on other events (*e.g.* from sensors) while waiting. The following simple program is an example of this. It lets the robot drive until either 10 seconds are past, or the touch sensor touches something.

```
task main()
{
   SetSensor(SENSOR_1,SENSOR_TOUCH);
   ClearTimer(3);
   OnFwd(OUT_A+OUT_C);
   until ((SENSOR_1 == 1) || (Timer(3) >100));
   Off(OUT_A+OUT_C);
}
```

Don't forget that timers work in ticks of 1/10 of a second, while *e.g.* the wait command uses ticks of 1/100 of a second.

The Display

It is possible to control the display of the RCX in two different ways. First of all, you can indicate what to display: the system clock, one of the sensors, or one the motors. This is equivalent to using the black view button on the RCX. To set the display type, use the command SelectDisplay(). The following program shows all seven possibilities, one after the other.

```
task main()
{
   SelectDisplay(DISPLAY_SENSOR_1); Wait(100); // Input 1
   SelectDisplay(DISPLAY_SENSOR_2); Wait(100); // Input 2
   SelectDisplay(DISPLAY_SENSOR_3); Wait(100); // Input 3
   SelectDisplay(DISPLAY_OUT_A); Wait(100); // Output A
   SelectDisplay(DISPLAY_OUT_B); Wait(100); // Output B
   SelectDisplay(DISPLAY_OUT_C); Wait(100); // Output C
   SelectDisplay(DISPLAY_WATCH); Wait(100); // System clock
}
```

Note that you should not use SelectDisplay(SENSOR_1).

The second way you can control the display is by controlling the value of the system clock. You can use this to display *e.g.* diagnostic information. For this use the command SetWatch(). Here is a tiny program that uses this:

```
task main()
{
   SetWatch(1,1); Wait(100);
   SetWatch(2,4); Wait(100);
   SetWatch(3,9); Wait(100);
   SetWatch(4,16); Wait(100);
   SetWatch(5,25); Wait(100);
}
```

Note that the arguments to SetWatch() must be constants.

Datalogging

The RCX can store values of variables, sensor readings, and timers, in a piece of memory called the datalog. The values in the datalog cannot be used inside the RCX, but they can be read by your computer. This is useful to *e.g.* check what is going on in your robot. Bricx Command Center has a special window in which you can view the current contents of the datalog.

Using the datalog consists of three steps: First, the NQC program must define the size of the datalog, using the command `CreateDatalog()`. This also clears the current contents of the datalog. Next, values can be written in the datalog using the command `AddToDatalog()`. The values will be written one after the other. (If you look at the display of the RCX you will see that one after the other, four parts of a disk appear. When the disk is complete, the datalog is full.) If the end of the datalog is reached, nothing happens. New values are no longer stored. The third step is to upload the datalog to the PC. For this, choose in Bricx Command Center the command **Datalog** in the **Tools** menu. Next press the button labelled **Upload Datalog**, and all the values appear. You can watch them or save them to a file to do something else with them. People have used this feature to *e.g.* make a scanner with the RCX.

Here is a simple example of a robot with a light sensor. The robot drives for 10 seconds, and five times a second the value of the light sensor is written into the datalog.

```
task main()
{
  SetSensor(SENSOR_2,SENSOR_LIGHT);
  OnFwd(OUT_A+OUT_C);
  CreateDatalog(50);
  repeat (50)
  {
    AddToDatalog(SENSOR_2);
    Wait(20);
  }
  Off(OUT_A+OUT_C);
}
```

Chapter 6

FPGA-BASED FUSED SMART SENSOR FOR DYNAMIC AND VIBRATION PARAMETER EXTRACTION IN INDUSTRIAL ROBOT LINKS

Carlos Rodriguez-Donate[1], Luis Morales-Velazquez[1], Roque Alfredo Osornio-Rios[1], Gilberto Herrera-Ruiz[2] and Rene de Jesus Romero-Troncoso[3,*]

[1] HSPdigital – CA Mecatronica, Facultad de Ingenieria, Universidad Autonoma de Queretaro, Campus San Juan del Rio, Rio Moctezuma 249, 76807 San Juan del Rio, Qro., Mexico; E-Mails: cdonate@hspdigital.org (C.R.-D.); lmorales@hspdigital.org (L.M.-V.); raosornio@hspdigital.org (R.A.O.-R.)

[2] Facultad de Ingenieria, Universidad Autonoma de Queretaro, Cerro de las Campanas s/n, 76010 Queretaro, Qro., Mexico; E-Mail: gherrera@uaq.mx

[3] HSPdigital – CA Telematica, DICIS, Universidad de Guanajuato, Carr. Salamanca-Valle km 3.5+1.8, Palo Blanco, 36700 Salamanca, Gto., Mexico

* Author to whom correspondence should be addressed; E-Mail: troncoso@hspdigital.org; Tel.: +52-464-647-9940; Fax: +52-464-647-9940.

ABSTRACT

Intelligent robotics demands the integration of smart sensors that allow the controller to efficiently measure physical quantities. Industrial manipulator robots require a constant monitoring of several parameters such as motion dynamics, inclination, and vibration. This work presents a novel smart sensor to estimate motion dynamics, inclination, and vibration parameters on industrial manipulator robot links based on two primary sensors: an encoder and a triaxial accelerometer. The proposed smart sensor implements a new methodology based on an oversampling technique, averaging decimation filters, FIR filters, finite differences and linear interpolation to estimate the interest parameters, which are computed online utilizing digital hardware signal processing based on field programmable gate arrays (FPGA).

Keywords

Smart sensor; motion dynamics; vibrations; accelerometer; FPGA.

1. INTRODUCTION

Intelligent robotics, as defined by Lopez-Juarez, *et al.* [1], demands the integration of smart sensors [2,3] that allow the controller to efficiently measure physical quantities. Communication and data processing functionalities are two of the most important features in smart sensors [3], but data fusion is also desirable. Industrial manipulator robots require constant monitoring of several variables and their fusion [4-6] such as: motion dynamics, inclination, and vibration; these variables inform about the machine wellness, highlighting the necessity of a specialized smart sensor that provides sufficient information to evaluate the robot performance. This work is focused on the extraction of several parameters from the mentioned physical variables, related to a single axis industrial robot arm. Motion dynamics is defined as the time-dependent profiles for position, velocity, acceleration, and jerk [7] in a servomotor, and determines the motion trajectory of a single axis robotic arm to reach a specific position and orientation based on a motion controller that uses these profiles as reference. On the other hand, the robot inclination is related to the spatial orientation of the physical sensor, where angular position, velocity, and acceleration on each robot link can be inferred. In addition, during the arm motion, vibrations are generated mainly due to friction, gearing, joint wear, *etc.*; these vibrations are undesired movements that reflect potential failures or improper operating conditions, making necessary their continuous monitoring to detect possible problems. Summarizing, it is desirable to have a single system able to provide all the aforementioned parameters from each robot link.

Current literature points out that the encoder in servomotors [4,8-10] and the accelerometer [11-14] are two of the most widely used sensors to monitor motion dynamics and vibrations on computerized numeric control (CNC) machines and robotic manipulator arms. Conversely, in industry, automation demands the integration of smart sensors [2,3] and control drivers in an open-architecture fashion [1,15]. Motion dynamics has been estimated from an incremental optical encoder in [9,10] where position, velocity, acceleration, and jerk parameters are successfully obtained, but they do not present the information of vibrations nor inclination. The use of accelerometers is well established to obtain kinematics parameters [8,16-18] or to measure vibrations [19], but there are no reported works that cover a broad parameter spectrum. The acceleration signal in an accelerometer contains merged information from the inclination with respect to gravity and about vibrations; therefore, a separation of these parameters is desirable for further kinematics and vibration monitoring. The extraction of vibrations allows failure detection [20] and some intelligent sensors have been proposed to perform this task [14]. Accelerometers have been also included in servo control loops [12].

By analyzing vibrations in combination with force sensors, contact forces are measured and calibrated [13], manifesting the relevance of a proper separation of vibration signals from the raw accelerometer measurement. Moreover, static acceleration indicates the inclination of the accelerometer with respect to gravity and by taking these signals, kinematics calibration in a manipulator arm can also be performed [11]. Inclination parameters have been investigated utilizing an accelerometer as primary sensor [21]. Furthermore, an integrated approach utilizing an encoder and an accelerometer has been presented to accurately estimate velocity [8]. Additionally, sensors are becoming more intelligent by integrating signal conditioning, processing units, communication protocols, among other features [3]. Therefore, the development of a smart sensor that integrates data fusion of motion dynamics, inclination, and vibration parameters is considered an essential move towards intelligent-robotics.

This work presents a novel smart sensor that extracts motion dynamics and inclination parameters along with the separation of vibration information from a single link in industrial robots, based on the fusion of two primary sensors: an optical incremental encoder and a triaxial accelerometer. Motion dynamics is estimated from the encoder measurement to give position, velocity, acceleration, and jerk; whereas vibrations and inclination are separated from the accelerometer signal, for providing angular position, velocity, acceleration, and vibrations. Estimated parameters are computed online utilizing digital hardware signal processing techniques such as digital filtering, interpolation, finite differences, among others. These computer-intensive processing algorithms are implemented in a field programmable gate array (FPGA) for a smart sensor approach by integrating hardware signal processing and data communication in an embedded system.

2. BACKGROUND

This section establishes the relationship among the estimated parameters on a robotics application, where the encoder gives information regarding motion dynamics and the accelerometer gives both inclination and vibration information.

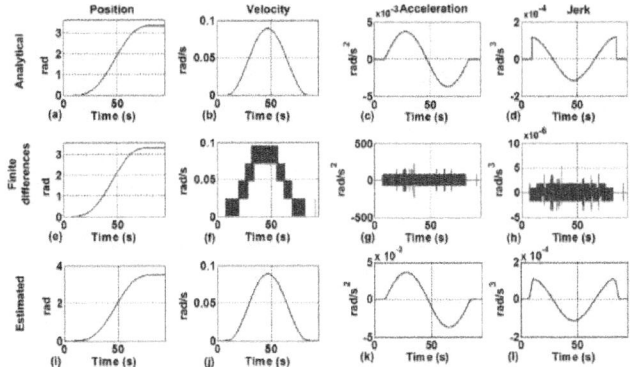

Figure 1. Analytical polynomial-motion profile: **(a)** position, **(b)** velocity, **(c)** acceleration, **(d)** jerk; recursive finite differentiation of the encoder: **(e)** position, **(f)** velocity, **(g)** acceleration, and **(h)** jerk; estimated motion profile with decimation: **(i)** position, **(j)** velocity, **(k)** acceleration, **(l)** jerk.

2.1. Motion Dynamics

Motion in a manipulator arm is conducted by the motion controller that applies a profile to perform smooth movements to the end effector. By taking the position feedback signal (p) from the servo control loop, velocity (v), acceleration (a), and jerk (j) can be estimated. Figure 1 shows an example of a polynomial profile comparing the analytical motion dynamics (Figure 1a, 1b, 1c, and 1d) against the motion dynamics obtained using recursive finite differentiation (Figure 1e, 1f, 1g, and 1h). In the case of using finite differentiation, quantization noise overwhelms the signal, making some filtering necessary. This differentiation and filtering stage can be performed by a combination of an averaging decimation filter [22], a finite difference stage, and a linear interpolation stage to estimate motion dynamic parameters $(p, v, a,$ and $j)$ as shown in Figure 1i, 1j, 1k, and 1l.

2.2. Inclination

From the accelerations provided by the accelerometer $(a_x, a_y,$ and $a_z)$, inclination angles pitch (ρ), roll (ϕ), and yaw (θ) are calculated. Figure 2 shows the diagram of a single axis robot arm depicting the triaxial accelerometer location, where ρ represents the inclination with respect to the X-axis, ϕ represents the inclination with respect to the Y-axis, and θ is the inclination with respect to the Z-axis.

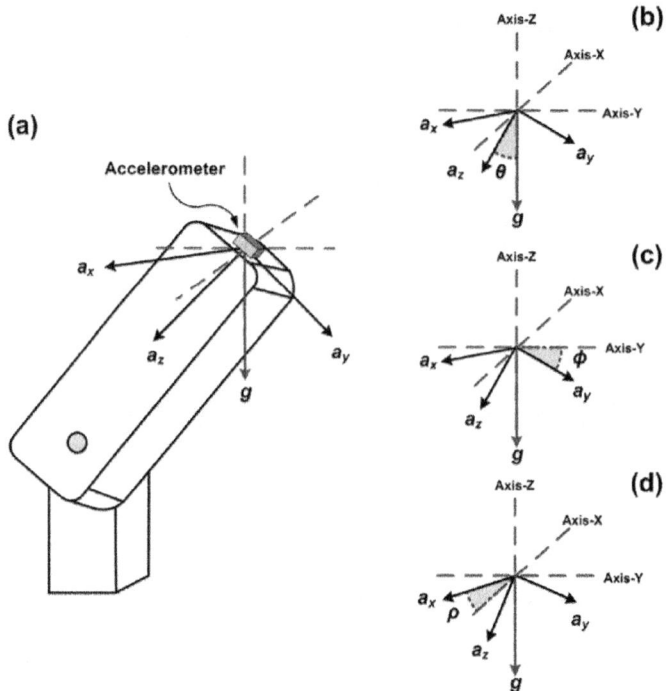

Figure 2. Triaxial accelerometer location on the robot arm, showing inclination angles with respect to gravity.

The inclination is calculated from the acceleration provided by the accelerometer as stated by (1-3) taken from [23]. These equations assume that original
signals (a_x, a_y, and a_z) are noise-free, which is unrealistic, requiring filtering.

$$\rho = \tan^{-1}\left(\frac{a_x}{\sqrt{a_y^2 + a_z^2}}\right) \tag{1}$$

$$\phi = \tan^{-1}\left(\frac{a_y}{\sqrt{a_x^2 + a_z^2}}\right) \tag{2}$$

$$\theta = \tan^{-1}\left(\frac{\sqrt{a_x^2 + a_y^2}}{a_z}\right) \tag{3}$$

2.3. Vibrations

The accelerometer provides merged information about vibration and inclination that must be separated; given that the inclination signal is principally low
frequency [24] whereas vibration is high frequency [25], they can be separated with
properly tuned filters. To reduce the noise and effectively extract the vibration
signal, a decimation process followed by a high-pass filter is used to separate the
three signals; this technique reduces the noise from the original signal.

3. SMART SENSOR METHODOLOGY

The proposed smart sensor is implemented in three stages: primary sensors,
signal conditioner, and signal processing; Figure 3 shows the overall architecture
for the proposed smart sensor. At the primary sensors stage the physical quantities
are sensed, at the signal conditioning stage the accelerometer signal is amplified,
and at the signal processing stage parameters are estimated. For each sampling
period, the encoder measures the relative servomotor shaft position and gives two
quadrature signals; the first step is to decode the absolute position of the servomotor shaft on the effector motions, then parameter estimation can be performed. The
accelerometer reads the acceleration analog signals and then the signal conditioner
amplifies these signals to pass them to the analog-to-digital converter, digitally
normalized (m/s^2). Once the sample is taken, the signal processing is started; at
the end of this process the estimated parameters are stored in the memory via the
memory interface; finally the user is able to request the desired information from
the smart sensor via the communication interface.

Most of the proposed smart sensor functionality is performed by the digital
signal processor inside the smart sensor, depicted in Figure 4. The motion dynamics block processes the encoder position signal *(E)* to obtain the motion dynamics
parameters *p*, *v*, *a*, and *j*. The inclination block takes the acceleration from each
axis (A_x, A_y and A_z) to calculate the inclination angles ρ, φ, and θ. The vibration
block separates the vibration signals from the inclination for further analysis. The
control unit synchronizes the resulting parameters to maintain the correspondence

among the signals; these parameters are then stored into an external synchronous dynamic random access memory (SDRAM) or sent via USB interface to the user PC.

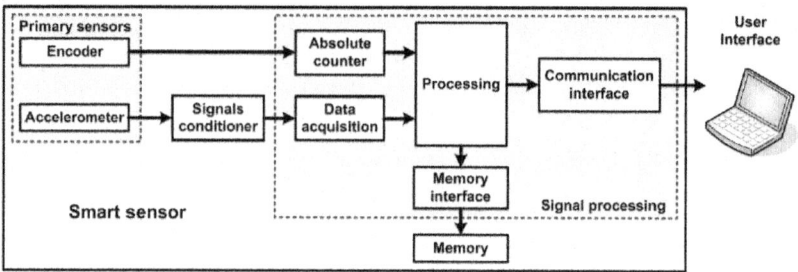

Figure 3. An overview of the architecture of the proposed smart sensor.

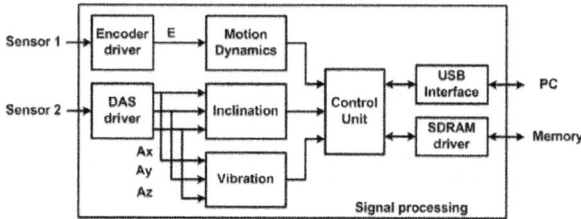

Figure 4. Smart sensor methodology showing the main blocks in the system.

The signal processing unit of the smart sensor needs to make intensive computations for obtaining motion dynamics, inclination, and vibration data from three acceleration signals and the encoder input. In order to properly achieve the required processing performance, an FPGA device is the most suited option for implementation thanks to the reconfigurability and parallelism features of these devices [25].

3.1. Motion Dynamics Methodology

The motion dynamics methodology is shown in Figure 5. Since the primary sensor in this module is the encoder, 64-times oversampling is applied, requiring a decimation factor of 64 that is also applied to a differentiation unit described below. This figure also shows the process used to estimate the motion profile using only the encoder signal by cascading three low noise differentiation units to obtain p, v, a, and j.

The averaging decimation filter shown in Figure 5 is used for filtering the encoder signal in order to reduce the quantization noise as stated in (4) and exposed by Rangel-Magdaleno *et al.* [22] using an oversampling technique, where E is the encoder input signal, p the decimated position, k the discrete-time index, and i the decimation index.

$$p\left(\frac{k}{64}\right) = \frac{1}{64}\sum_{i=0}^{63} E(k-i)$$

$$(4)$$

The differentiation unit performs a low-noise difference over p utilizing a finite difference operation defined in (5) to obtain p_F. An averaging decimation filter is then applied to the differentiated signal as stated in (6) to obtain p_D, and a linear interpolation using (7) to preserve the sampling rate and obtain the estimated velocity v. To estimate a and j an equivalent process is followed, recursively.

$$p_F(k) = p(k) - p(k-1) \tag{5}$$

$$p_D\left(\frac{k}{64}\right) = \frac{1}{64}\sum_{i=0}^{63} p_F(k-i) \tag{6}$$

$$v(64 \cdot k + i) = p_D(k-1) + \frac{1}{64}[p_D(k) - p_D(k-1)] \cdot i; \quad i = 0,1,...63 \tag{7}$$

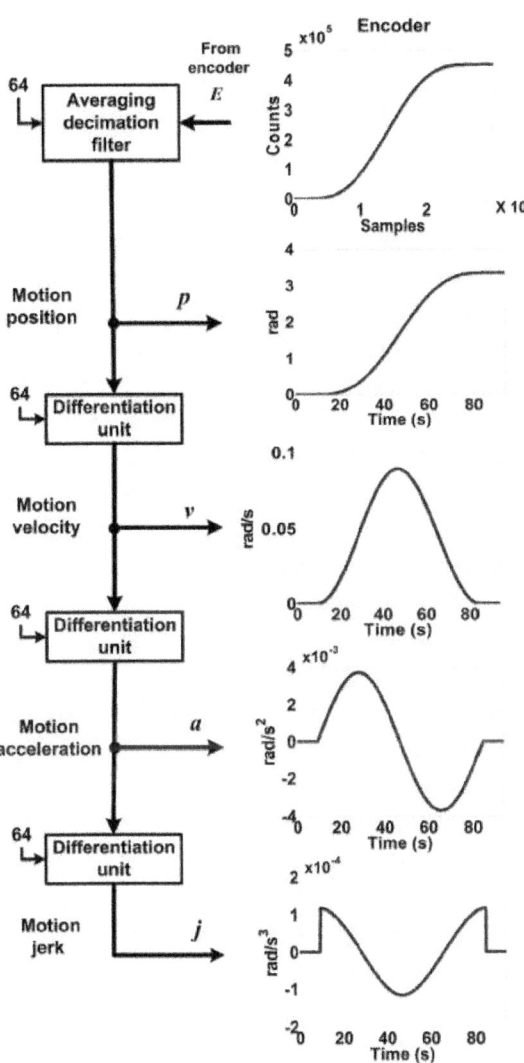

Figure 5. The motion dynamics methodology.

3.2. Inclination Methodology

The process for estimating inclination parameters is depicted in Figure 6. The raw triaxial acceleration signals are filtered by averaging decimation filters of the 128th order, as stated in (8), where subscript s represents the accelerometer axis (x, y, or z). Then, to separate inclination signals from vibration, a 32nd order low-pass FIR filter is applied to the decimated signals resulting in the inclination signals (A_{xdf}, A_{ydf} and A_{zdf}). To calculate inclination angles from the acceleration signals, it is necessary to compute the inverse tangent function; this is performed using the CORDIC algorithm [27].

$$A_{sd}\left(\frac{k}{128}\right) = \frac{1}{128}\sum_{i=0}^{127} A_s(k-i) \tag{8}$$

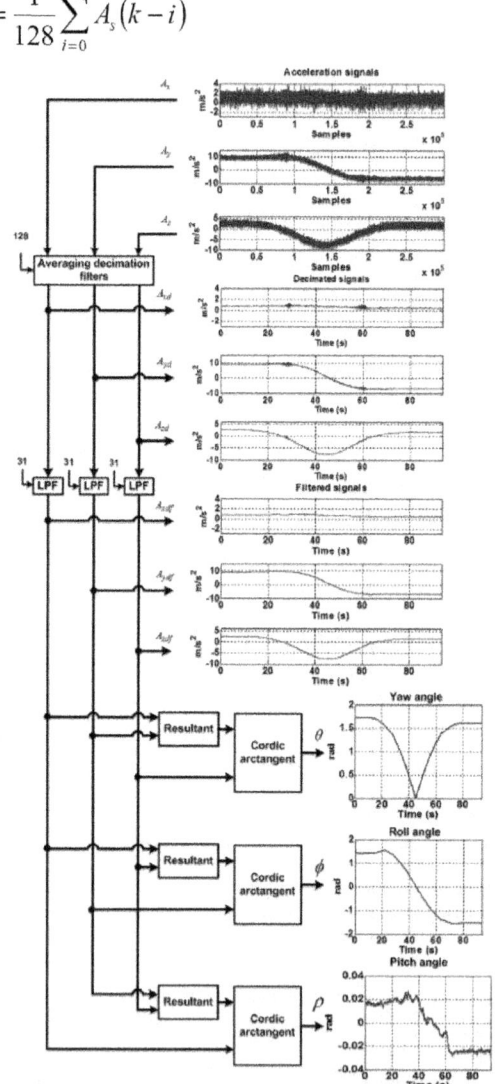

Figure 6. Methodology to estimate inclination angles from accelerometer signals.

3.3. Vibration Methodology

To extract the vibration components from the raw accelerometer signals they are first filtered using a 32nd order averaging decimated filter as depicted in Figure 7; after this, the decimated signals (A_{xd}, A_{yd} and A_{zd}) are passed through a 1024th order high-pass FIR filter to isolate the vibration signals (V_x, V_y and V_z).

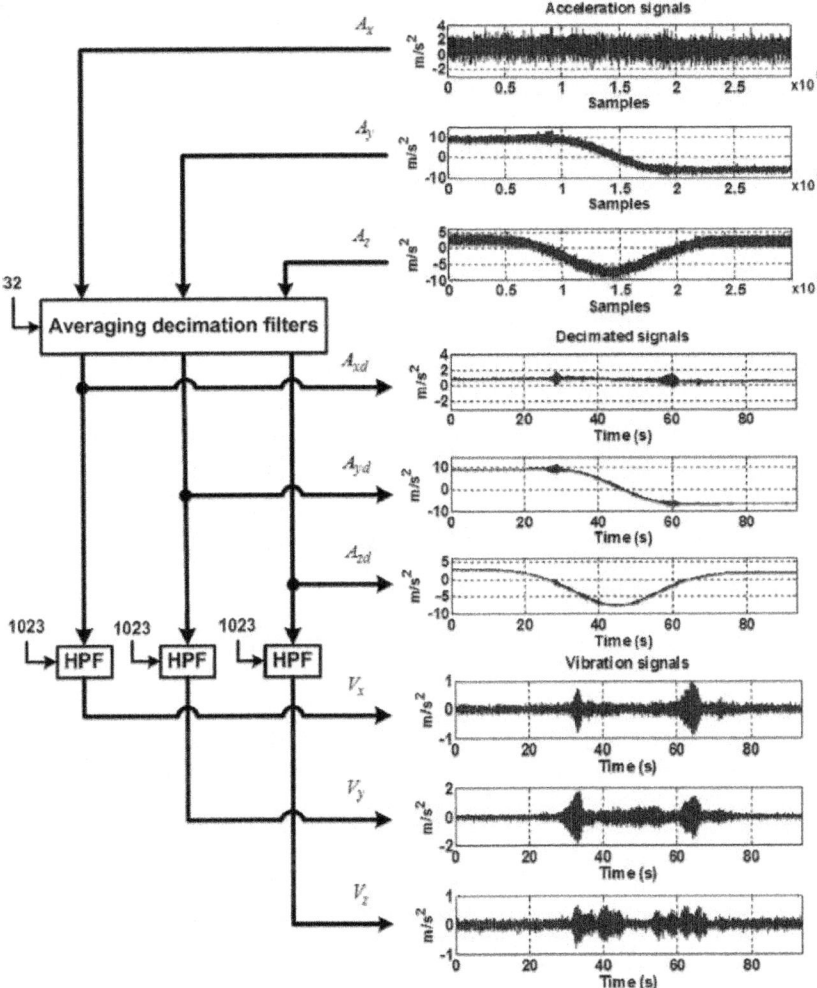

Figure 7. Vibration-parameters estimation methodology used to separate vibrations from the original accelerometer measurement.

4. EXPERIMENTAL RESULTS

In this section, the experimental validation of the proposed smart sensor is presented. The online parameter estimation was performed during a single arm movement in the second effector.

4.1. Experimental Setup

The experimental setup consists on the instrumentation of a single axis of a 6-degree of freedom Cloos-Romat 56 modular robot, a triaxial accelerometer LIS3L02AS4 [28] with a signal bandwidth of 750 Hz, and a sampling frequency of 3.2 kHz at the three-channel on-board data acquisition system; a proprietary Spartan 3E XC3S1600E FPGA platform running at 48 MHz as the smart sensor processing unit, a proprietary motion controller, and a user PC, as shown in Figure 8. Processing units were implemented in the VHSIC hardware description language (VHDL) under the Xilinx ISE Design Suite version 11, since it has being extensively used to develop smart sensors [29,30]. Table 1 summarizes the resource usage of the FPGA after compilation.

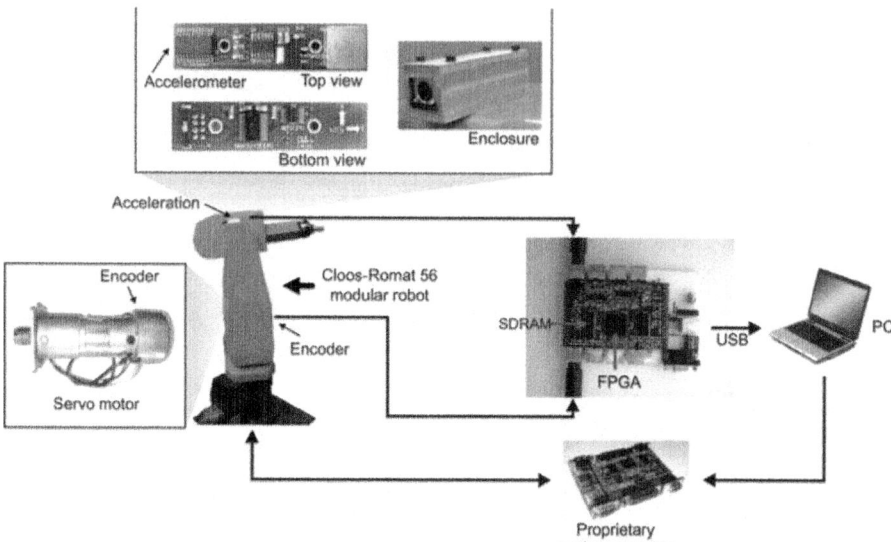

Figure 8. Experimental setup used for experimentation, showing the Cloos-Romat 56 modular robot, the two primary sensors, the FPGA processing unit, the proprietary motion controller, and the user PC.

Table 1. FPGA resource usage.

1.6 million-gate Xilinx Spartan 3E FPGA: XC3S1600E			
Element	Used	Available	Percentage %
Slices	2027	4752	13
Slice Flip-flops	2158	2950	47
4-input LUTs	2719	2959	49
Block RAMs	6	36	17
Multipliers	25	36	70

The experiment consists of a single-axis movement of the second effector arm using the proprietary motion controller by applying a 7th order polynomial motion profile [7] during 90 seconds and extracting in real-time the required parameters that are stored in SDRAM and then sent to the user PC.

4.2. Motion Dynamics Results

Figure 9 shows the analytical profiles that were obtained from the motion controller, whereas the estimated ones are computed by the smart sensor for the motion dynamics estimation. The upper row is the profile applied to the motion controller to conduct the arm movement; the middle row is the estimated profile obtained from the encoder measurement by applying the proposed methodology, and the bottom row is the computed error between these signals. For all parameters the shape of the analytical profile is well fitted, and the error between the analytical and the estimated signals is quantitatively calculated. The estimated position profile in Figure 9e is very similar to the analytical one in Figure 9a, having an absolute error below 0.01%. The estimated velocity in Figure 9f fits the analytical profile of Figure 9b with an overall absolute error below 0.5%. The estimated acceleration of Figure 9g is very similar in shape to the analytical one in Figure 9c with an absolute error below 5%. The estimated jerk on Figure 9h also shows a good fitting to the analytical one in Figure 9d, but the error is increased mainly because the initial and final discontinuities in the jerk shape, having an absolute error below 20% in those sections and below 5% in the middle section.

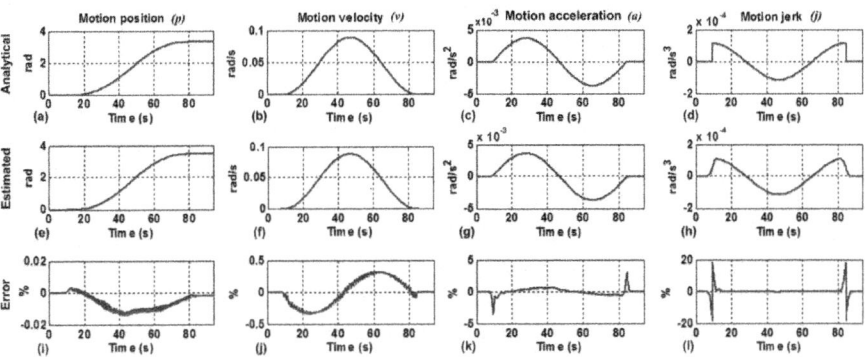

Figure 9. Motion dynamics results: **(a)** analytical position profile, **(e)** estimated position profile, **(i)** position error, **(b)** analytical velocity profile, **(f)** estimated velocity profile, **(j)** velocity error, **(c)** analytical acceleration profile, **(g)** estimated acceleration profile, (k) acceleration error, **(d)** analytical jerk profile, **(h)** estimated jerk profile, and **(l)** jerk error.

4.3. Inclination Results

The inclination results from the accelerometer measurements follow the methodology described in section 3.2. The estimated inclination angles shown in Figure 10 were computed using the same trajectory used in the previous section for motion dynamics. Analytical inclination angles were calculated from the motion position profile. The analytical yaw in Figure 10a, compared with the estimated yaw in Figure 10d, gives a maximum error of 6% in Figure 10g. Analytical roll of Figure 10b is similar to the estimated roll of Figure 10e with an error below 10% on Figure 10h. Finally, analytical pitch angle of Figure 10c, compared with the estimated pitch of Figure 10f, gives an error below 1% on Figure 10i. As presented in Figure 10, analytical and estimated angles are very similar in shape with an overall error below 10%.

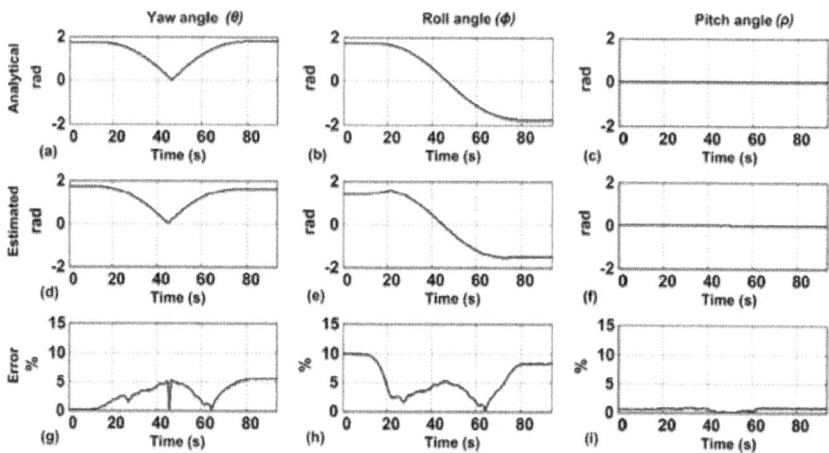

Figure 10. Inclination estimated results: (a) analytical yaw, (b) analytical roll, (c) analytical pitch, (d) estimated yaw, (e) estimated roll, (f) estimated pitch, (g) yaw relative error (h) roll relative error, and (i) pitch relative error.

4.4. Vibration Results

The vibration separation performed by the smart sensor is shown in Figure 11; this figure shows the three axis vibration signals, separated from the accelerometer measurements. The obtained vibration signals are identified with respect to time and the motion profile shape because they were taken using the same primary sensor in the system. Hence it is possible to relate the inclination results with vibration results and, in a further analysis, to identify gearing or friction problems.

4.5. Discussion

Experimental results using the proposed smart sensor present the obtained motion dynamics, inclination angles and vibrations over a single axis robot link. In the case of motion dynamics, position, velocity, and acceleration were obtained having an overall error below 5%, and below 20% for jerk estimation. Inclination angles were successfully separated from vibrations and the error between the analytical and the estimated values were below 10% for all angles. The separated vibration signals from the smart sensor do not contain information on the inclination and the user can further process the information for monitoring and diagnosis purposes.

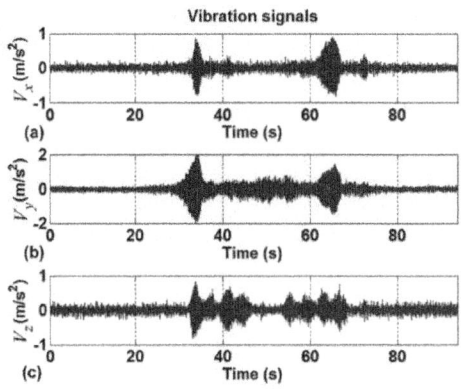

Figure 11. Vibration information separated from inclination at the accelerometer
(a) X-axis, (b) Y-axis, and (c) Z-axis.

The parameter extraction is possible in the developed smart sensor thanks to the FPGA parallelism and reconfigurability. These features allow the hardware-processing unit of the smart sensor to efficiently perform the related smart operations such as hardware signal processing (data acquisition drivers, CORDIC arc tangent estimation, and oversampling, decimation, FIR, and interpolation filtering), data storage, and data communication. The developed sensor can be utilized in many different areas with a variety of applications; Table 2 presents several application examples on high-impact researches where the developed smart sensor can be utilized.

Table 2. Smart sensor parameter application areas.

Primary sensor	Parameter group	Parameter	Applications
Encoder	Motion dynamics	Position	Positioning [10,31]
		Velocity	Control loop [32,33]
		Acceleration	Control loop [8]
		Jerk	Machine wear [7,31]
Accelerometer	Inclination	Angular position	Calibration [11], positioning [16,18]
		Angular velocity	Control loop [17,18]
		Angular acceleration	Torque control [12,13]
	Vibrations	Vibration	Failure detection [14]

5. CONCLUSIONS

This work proposes a new smart sensor to simultaneously obtain several parameters related to motion dynamics and inclination, along with the separation of the vibration information using two primary sensors: an encoder and a triaxial accelerometer on a single link of industrial robots. Results on motion dynamics and inclination estimations show the effectiveness of this smart sensor that integrates data fusion among primary sensor data. The vibration signal result, provided by the smart sensor, does not contain the inclination information and it contains the vibration information only, which can be further processed for monitoring and diagnosis purposes. Furthermore, the proposed smart sensor was implemented in a low-cost FPGA where hardware signal processing units compute in parallel the parameters and integrate the necessary modules with a system on-a-chip approach. The developed sensor provides the information of a single robot link, but one smart sensor can be placed on each robot link and combine the obtained information from all of them to estimate other parameters such as direct kinematics or make assessments on the robot wellness; however, this is beyond the scope of the present research and it is left for future work.

Acknowledgements

The authors wish to thank the reviewers for having made suggestions and comments that greatly improved the paper. This project was partially supported by CONACyT scholarship 217623, FOMIX-QRO-2008-CO2-102123 and SEP-CONACyT 84723 projects.

REFERENCES

1. Lopez-Juarez, I.; Corona-Castuera, J.; Peña-Cabrera, M.; Ordaz-Hernandez, K. On the design of intelligent robotic agents for assembly. *Inf. Sci.* **2005**, *171*, 377-402.

2. Hernandez, W. A survey on optimal signal processing techniques applied to improve the performance of mechanical sensors in automotive applications. *Sensors* **2007**, *7*, 84-102.

3. Rivera, J.; Herrera, G.; Chacón, M.; Acosta, P.; Carrillo, M. Improved progressive polynomial algorithm for self-adjustment and optimal response in intelligent sensors. *Sensors* **2008**, *8*, 7410-7427.

4. Olabi, A.; Béarée, R.; Gibaru, O.; Damak, M. Feedrate planning for machining with industrial six-axis robots. *Control Eng. Pract.* **2010**, in press, doi:10.1016/j.conengprac.2010.01.004.

5. Zhu, W.H.; Lamarche, T. Velocity estimation by using position and acceleration sensors. *IEEE T. Ind. Electron.* **2007**, *54*, 2706-2715.

6. Jeon, S.; Tomizuka, M. Benefits of acceleration measurement in velocity estimation and motion control. *Control Eng. Pract.* **2007**, *15*, 325-332.

7. Osornio-Rios, R.A.; Romero-Troncoso, R.J.; Herrera-Ruiz, G.; Castañeda-Miranda, R. FPGA implementation of higher degree polynomial acceleration profiles for peak jerk reduction in servomotors. *Robot. Cim-Int. Manuf.* **2009**, *25*, 379-392.

8. Väliviita, S.; Ovaska, S.J. Delayless acceleration measurement method for elevator control. *IEEE T. Ind. Electron.* **1998**, *45*, 364-366.

9. de Santiago-Pérez, J.J.; Osornio-Rios, R.A.; Romero-Troncoso, R.J.; Herrera-Ruiz G.; Delgado-Rosas, M. DSP algorithm for the extraction of dynamics parameters in CNC machine tool servomechanisms from an optical incremental encoder. *Int. J. Mach. Tool Manu.* **2008**, *48*, 1318-1334.

10. Morales-Velazquez, L.; Romero-Troncoso, R.J.; Osornio-Rios, R.A.; Cabal-Yepez E. Sensorless jerk monitoring using an adaptive antisymmetric high-order FIR filter. *Mech. Syst. Signal Pr.* **2009**, *23*, 2383-2394.

11. Canepa, G.; Hollerbach, J.M.; Boelen, A.J.M.A. Kinematic calibration by means of a triaxial accelerometer. In *Proceedings of ICRA-1994, the International Conference on Robotics and Automation*, San Diego, CA, USA, May 1994.

12. Han, J.D; He, Y.Q.; Xu, W.L. Angular acceleration estimation and feedback control: an experimental investigation. *Mechatronics* **2007**, *17*, 524-532.

13. Gamez-Garcia, J.; Robertsson, A.; Gomez-Ortega, J; Johansson, R. Self-calibrated robotic manipulator force observer. *Robot. Cim-Int. Manuf.* **2009**, *25*, 366-378.

14. Wang, W.; Jianu, O.A. A smart sensing unit for vibration measurement and monitoring. *IEEE/ASME Trans. Mechatron.* **2010**, *15*, 70-78.

15. Mekid, S.; Pruschek, P.; Hernandez, J. Beyond intelligent manufacturing: a new generation of flexible intelligent NC machines. *Mech. Mach. Theory* **2009**, *44*, 466-476.

16. Liu, H.H.S.; Pang, G.K.H. Accelerometer for mobile robot positioning. *IEEE T. Ind. Appl.* **2001**, *37*, 812-819.

17. Jassemi-Zargani, R.; Necsulescu, D. Extended kalman filter-based sensor fusion for operational space control of a robot arm. *IEEE T. Instrum. Meas.* **2002**, *51*, 1279-1282.

18. Dumetz, E.; Dieulot, J.Y.; Barre, P.J.; Colas, F.; Delplace, T. Control of an industrial robot using acceleration feedback. *J. Intell. Robot Syst.* **2006**, *46*, 111-128.

19. Ohta, T.; Murakami, T. A stabilization control of bilateral system with time delay by vibration index-application to inverted pendulum control. *IEEE T. Ind. Electron.* **2009**, *56*, 1595-1603.

20. Trendafilova, I.; van Brussel, H. Condition monitoring of robot joints using statistical and nonlinear dynamics tools. *Meccanica* **2003**, *38*, 283-295.

21. Bernmark, E.; Wiktorin, C. A triaxial accelerometer for measuring arm movements. *Appl. Ergon.* **2002**, *33*, 541-547.

22. Rangel-Magdaleno, J.J.; Romero-Troncoso, R.J.; Osornio-Rios, R.A.; Cabal-Yepez, E. Novel oversampling technique for improving signal-to-quantization noise ratio on accelerometer-based smart jerk sensors in CNC applications. *Sensors* **2009**, *9*, 3767-3789.

23. AN3461 Application Note, Tilt sensing using linear accelerometers. *Freescale Semiconductor*, 2007.

24. Miro, J.V.; White, A.S. Modelling an industrial manipulator a case study. *Simulat. Pract. Theor.* **2002**, *9*, 293-319.

25. Karagülle, H.; Malgaca, L. Analysis of end point vibrations of a two-link manipulator by Integrated CAD/CAE procedures. *Finite Elem. Anal. Des.* **2004**, *40*, 2049-2061.

26. Sulaiman, N.; Obaid, Z.A.; Marhaban M.H. and Hamidon M.N. Design and Implementation of FPGA-Based Systems - A Review. *Aust. J. Basic Appl. Sci.* **2009**, *3*, 3575-3596.

27. Vachhani, L.; Sridharan, K.; Meher, P.K. Efficient FPGA realization of CORDIC with application to robotic exploration. *IEEE T. Ind. Electron.* **2009**, *56*, 4915-4929.

28. LIS3L02AS4 data sheet, STMicroelectronics, 2004.

29. Patra, J.C.; Lee, H.Y.; Meher, P.K.; Ang, E.L. Field programmable gate array implementation of a neural network-based intelligent sensor system. In *International Conference on Control Automation Robotics and Vision (IEEE ICARCV 2006)*, Singapore, December 2006, pp. 333-337.

30. Depari, A.; Ferrari, P.; Flammini, A.; Marioli, D.; Taroni, A. A VHDL model of a IEEE1451.2 smart sensor: characterization and applications. *IEEE Sens. J.* **2007**, *7*, 619-626.

31. Altintas, Y. *Manufacturing Automation: Metal Cutting Mechanics, Machine Tool Vibrations, and CNC Design*; Cambridge University Press: Cambridge, UK, 2000.

32. Liu, G; Goldenberg, A.A.; Zhang, Y. Precise slow motion control of a direct-drive robot arm with velocity estimation and friction compensation. *Mechatronics* **2004**, *14*, 821-834.

33. Fan, S; Bicker, R. Design and validation of an FPGA-based self-healing controller for hybrid machine tools. *J. Adv. Mech. Sys.* **2010**, *2*, 99-107.

This page left intentionally blank.

INDEX